Essentials of

Anatomy & Physiology

LABORATORY MANUAL

Bert Atsma

Leigh Rapaport Levitt

Richard E. McKeeby

Jessica Potell Sand

Union County College
Cranford, New Jersey

An imprint of Addison Wesley Longman, Inc.

Menlo Park, California • Reading, Massachusetts • New York • Harlow, England
Don Mills, Ontario • Sydney • Mexico City • Madrid • Amsterdam

Executive Editor: Daryl Fox
Associate Editor: Claire Brassert
Managing Editor: Wendy Earl
Production Editor: David Novak
Cover and Text Design: Brad Greene, Greene Design

The authors and publisher believe that the laboratory experiments described in this publication, when conducted in conformity with the safety precautions described herein and according to the school's laboratory safety procedures, are reasonably safe for the students for whom this manual is directed. Nonetheless, many of the described experiments are accompanied by some degree of risk, including human error, the failure or misuse of laboratory or electrical equipment, mismeasurement, spills of chemicals, and exposure to sharp objects, heat, bodily fluids, blood, or other biologics. The author and publisher disclaim any liability arising from such risks in connection with any of the experiments contained in this manual. If students have questions or problems with materials, procedures, or instructions on any experiment, they should always ask their instructor for help before proceeding.

Library of Congress Cataloging-in-Publication Data

Essentials of anatomy and physiology laboratory manual / Bert Atsma
. . . [et al.].
p. cm.
Includes index.
ISBN 0-8053-5092-6
1. Human physiology—Laboratory manuals. 2. Human anatomy—Laboratory manuals. I. Atsma, Bert.
QP44.E83 1999
612'.0078—dc21 98-46568
CIP

ISBN 0-8053-5092-6

3 4 5 6 7 8 9 10–CRS–02 01 00

2725 Sand Hill Road
Menlo Park, California 94025

Table of Contents

About the Authors

Bert Atsma

Bert Atsma teaches lectures and labs in Human Anatomy and Physiology, Introduction to Biology, and Human Biology at Union County College. A former regulatory toxicologist, he has lectured on epidemiology, ecology, general biology, and genetics. He earned his B.S. and M.S. degrees from Montclair State University.

Richard E. McKeeby

Richard McKeeby teaches lectures and labs in Introduction to Biology and Human Biology at Union County College. Previously, he has taught Vertebrate and Invertebrate Zoology, Animal Biology, and New Jersey Natural History. He earned his B.S. degree from Rutgers University, and his M.S. was awarded by Union College.

Leigh Rapaport Levitt

Leigh Levitt ia a laboratory science instructor at Union County College, currently teaching Human Anatomy and Physiology. She earned a B.S. at the Univesity of Connecticut College of Pharmacy and is registered pharmacist.

Jessica Potell Sand

Jessica Sand teaches lectures in Anatomy and Physiology, Microbiology, Human Biology, Nutrition, and Pathology at Union County College. She earned a B.S. and an M.A. in Biology from Brooklyn College (CUNY) and M.S. in Microbiology and Public Health from Wagner College. She is ABD in Cell Biology from CUNY and has research experience in the area of neuroimmunology.

Preface

We dedicate this lab manual to all of the teachers who played a role in our formal training and prepared us to pass on to a new generation our interest in and our love of science.

We would like to thank our colleagues, past and present, and all who have helped us through the process that culminated in this lab manual. Special thanks are extended to: William Dunscombe, Helen Gmitro, Jane Healy-McMillin, Judy Kushnick, Phyllis Mayer, Marcia Meyers, Liz Patberg, and Irene Williams.

We would like to thank the following reviewers for their helpful insights:

Victor Eroschenko, University of Idaho

Henriette Evans, Pennsylvania College of Technology

Patrick Fulks, Bakersfield College

Daniel Gong, Seattle Central Community College

Laura Hebert, Angelina College

Paul Jarrell, Pasadena City College

Robert Keck, Indiana Vocational Technical Ivy College

Steve Kirk, Parker College

Mary Morgan, Fort Hays State University

Charles Page, El Camino College

Michael Patrick, Pennsylvania State University

Eugene Rutheny, Westchester Community College

Leba Sarkis, Aims Community College

Rhonda Shepperd, College of West Virginia

Mary Lynne Stephanou, Santa Monica Community College

Patricia Turner, Howard Community College

Margaret Voss, Syracuse University

We would also like to thank Claire Brassert, at Benjamin/Cummings, and David Novak, at Wendy Earl Productions, for their tireless work on the manual.

Finally, we would like to thank our families for their support and encouragement throughout this labor of love.

Note to Instructors: Answers to the questions in the lab manual can be found online at www.awl.com/bc. At the Benjamin/Cummings Science site, click Product Catalog in the list on the left. In the Product Catalog window, click Author Index in the list on the left. A list of products appears, sorted by author. Scroll to Atsma, and click Essentials of Anatomy & Physiology Lab Manual. At the Essentials of Anatomy & Physiology Lab Manual page on the right side, under For Instructors Only, click Lab Manual Answer Key, and follow the instructions for viewing the answers.

This is a password-protected site. Please contact your Benjamin/Cummings sales representative for the password.

Photo Credits

Figure 1.1 Courtesy of Leica Microsystems, Inc.

Figure 2.4; 4.2; 4.3; 4.6; 6.1 © Ed Reschke

Figure 4.1; 4.4; 4.5 © Biophoto/Photo Researchers, Inc.

Figure 4.7 © Andrew Kuntzman

Art Credits

Shirley Bortoli: Figure 1.2; 1.3; 1.4; 4.4; 4.6; 4.7; 7.1; 11.2; 11.3

Charles Bridgeman: Figure 10.2
From *Principles of Anatomy and Physiology,* 8e; by Gerard J. Tortora and Sandra R. Grabowski, © 1996 by Addison-Wesley Educational Publishers. Reprinted by permission.

Todd A. Buck: Figure 10.4; 10.5
From *Health Assessment in Nursing,* 1e; by Linda Sims, Donita D'Amico, Johanna Stiesmeyer, and Judith A. Webster, © 1995 by The Benjamin/Cummings Publishing Company. Reprinted by permission.

Christopher Burke: Figure 13.2
From *Fundamentals of Nursing,* 5e; by Barbara Kozier, Glenora Erb, Kathleen Blais, and Judith Wilkinson, © 1995 by The Benjamin/Cummings Publishing Company. Reprinted by permission.

Barbara Cousins: Figure 6.1b
From *Health Assessment in Nursing,* 1e; by Linda Sims, Donita D'Amico, Johanna Stiesmeyer, and Judith A. Webster, © 1995 by The Benjamin/Cummings Publishing Company. Reprinted by permission.

Leonard Dank: Figure 6.1; 6.2; 6.3; 6.4; 6.5; 6.6a-c; 6.7; 6.8; 6.9; 6.10a-d; 7.4; 7.5a,b; 8.9
From *Principles of Anatomy and Physiology,* 8e; by Gerard J. Tortora and Sandra R. Grabowski, © 1996 by Addison-Wesley Educational Publishers. Reprinted by permission.

Sharon Ellis: Figure 8.3; 8.4; 8.6; 8.7; 8.8; 9.4; 9.2b,c; 9.5
From *Principles of Anatomy and Physiology,* 8e: by Gerard J. Tortora and Sandra R. Grabowski, © 1996 by Addison-Wesley Educational Publishers. Reprinted by permission.

Nea Hanscomb: Figure 8.1; 9.2a; 9.3; 11.4; 13.2; 13.3a,b; 13.6; 15.5

Lauren Kenswick: Figure 2.1; 4.5; 5.1
From *Principles of Anatomy and Physiology,* 8e; by Gerard J. Tortora and Sandra R. Grabowski, © 1996 by Addison-Wesley Educational Publishers. Reprinted by permission.

Romaine LoPrete: Figure 9.1
From *Health Assessment in Nursing,* 1e; by Linda Sims, Donita D'Amico, Johanna Stiesmeyer, and Judith A. Webster, © 1995 by The Benjamin/Cummings Publishing Company. Reprinted by permission.

Biagio John Melloni: Figure 10.1
From *Principles of Anatomy and Physiology,* 8e; by Gerard J. Tortora and Sandra R. Grabowski, © 1996 by Addison-Wesley Educational Publishers. Reprinted by permission.

Kristin N. Mount: Figure 8.5
From *Health Assessment in Nursing,* 1e; by Linda Sims, Donita D'Amico, Johanna Stiesmeyer, and Judith A. Webster, © 1995 by The Benjamin/Cummings Publishing Company. Reprinted by permission.

Hilda Muinos: Figure 7.2; 9.6; 13.1; 13.4; 13.5; 13.7; 13.8; 14.2
From *Principles of Anatomy and Physiology,* 8e; by Gerard J. Tortora and Sandra R. Grabowski, © 1996 by Addison-Wesley Educational Publishers. Reprinted by permission.

Lynn O'Kelley: Figure 3.2; 14.1; 14.2; 14.3
From *Principles of Anatomy and Physiology,* 8e; by Gerard J. Tortora and Sandra R. Grabowski, © 1996 by Addison-Wesley Educational Publishers. Reprinted by permission.

Dr. Joel Schechter: Figure 3.5; 3.6; 15.3a,b; 16.3; 17.5; 18.5
From *Laboratory Manual for Anatomy and Physiology, with Fetal Pig Dissections,* 2e; by Patricia J. Donnelly and George A. Wistreich, © 1997 by The Addison-Wesley Educational Publishers. Reprinted by permission.

Nadine Sokol: Figure 2.5; 4.1a; 4.2a; 4.3a; 8.2; 10.3; 12.1; 15.1; 16.2
From *Principles of Anatomy and Physiology,* 8e; by Gerard J. Tortora and Sandra R. Grabowski, © 1996 by Addison-Wesley Educational Publishers. Reprinted by permission.

Kevin Somerville: Figure 2.2; 3.1; 3.4; 3.7; 14.2; 16.1a,b; 17.2; 17.3; 17.4; 18.2; 18.3; 18.4
From *Principles of Anatomy and Physiology,* 8e; by Gerard J. Tortora and Sandra R. Grabowski, © 1996 by Addison-Wesley Educational Publishers. Reprinted by permission.

Beth Willert: Figure 7.2; 7.3
From *Principles of Anatomy and Physiology,* 8e; by Gerard J. Tortora and Sandra R. Grabowski, © 1996 by Addison-Wesley Educational Publishers. Reprinted by permission.

Figure 14.4; 17.1
From *Biology: Concepts and Connections,* 2e; by Paul Campbell, Larry Mitchell, and Jane Reece, © 1997 by Addison Wesley Longman, Inc.

EXERCISE 1

The Microscope

OBJECTIVES

After completing this exercise, you should be able to:

1. Identify the main parts of a microscope and explain their basic functions.
2. Effectively use a microscope to view both prepared slides and wet mounts.
3. Care for and clean the parts of a microscope that affect its performance.
4. Understand the concepts of magnification, size of field, depth of field, and focusing.

MATERIALS

- ❑ Prepared slides of the letter "e"
- ❑ Prepared slides of contrasting color threads
- ❑ Plain microscope slides
- ❑ Cover slips

Dropper bottles containing:

- ❑ Isotonic saline solution
- ❑ Iodine solution
- ❑ Lens cleaning solution

- ❑ Toothpicks
- ❑ Lens paper

Introduction

Many years ago, scientists discovered that a system using two lenses could yield far greater magnification than could single-lens magnifying devices. Your microscopes, while much easier to use than those of the 19th century, apply this same basic 2-lens principle. The major innovations in your microscopes lie in the focusing controls and the stage and light adjustment devices.

There are basically two different styles of microscopes, monocular (one ocular) and binocular (two ocular). Although easier on the eyes over the long term, binocular scopes can be tricky to use at first. Monocular scopes are less expensive, yet offer the same magnification and resolution quality as binocular microscopes. Whichever your laboratory has, a little practice will go a long way in helping you become comfortable with the microscope.

Anatomy of the Microscope

Before you can effectively use a microscope, it is important to learn its parts and basic functions. Consult Figure 1.1 as you read this section.

The **ocular** is the eyepiece lens, usually 10X, located at the top of the microscope's body tube. At the other end of the body tube is the **revolving nosepiece**, on which are attached the **objective lenses**. You turn the nosepiece in order to view your specimen with different objective lenses, which offer different magnification conditions. Microscopes vary

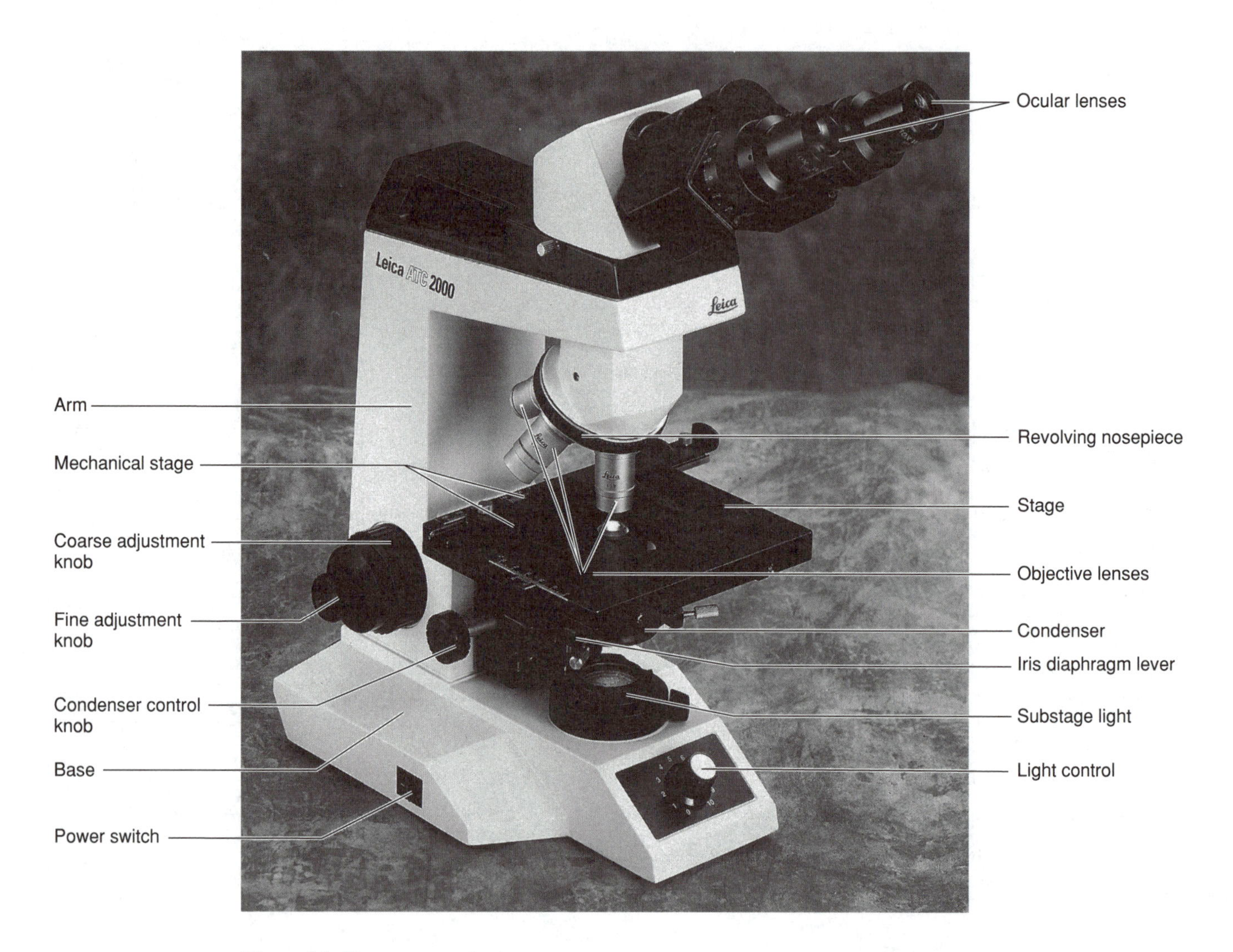

Figure 1.1 The compound microscope.

greatly in their selection of objective lenses. Some microscopes have the very powerful 100X lens, or **oil immersion lens**, so called because a drop of immersion oil is often necessary to help clarify the image. The highest-power lens below this oil immersion lens is called the **high-dry objective**, usually a 40–45X lens (Figure 1.2).

Attached to the **stage**, or the platform on which your slide will be placed, are a variety of tools. There is always some kind of **clamp** to hold the slide, and the better microscopes will also have a device to move the clamp, the stage, or both (and thus the slide as well). The most popular version of this device is the **mechanical stage control**. Beneath the stage is a **condenser** with an **iris diaphragm** underneath it. The condenser and iris diaphragm channel and regulate the light passing through your slide.

The **focusing knobs** adjust the distance between your slide and the objective lens by moving either the body tube or the stage up and down. It is important to learn the difference between the **coarse** and **fine adjustment** focusing knobs, as you can damage a slide or objective lens by using the wrong knob to focus.

The **arm** and **base** make up most of the rest of the microscope's frame, and are good places to hold the microscope when carrying it.

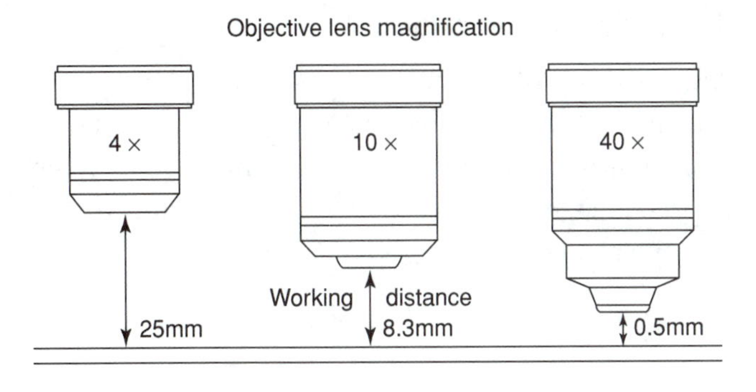

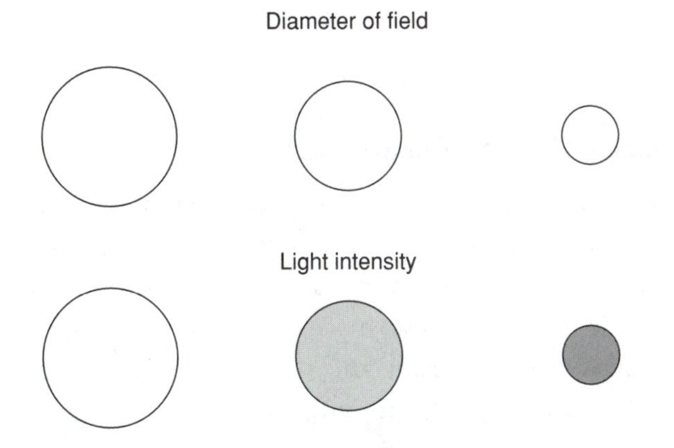

Figure 1.2 The relationship between the working distance of the low (4X), medium (10X), and high-dry (40–45X), objectives and the iris diaphragm adjustments.

Magnification Issues

To calculate the total **magnification** of the object you are viewing, simply multiply the ocular magnification (usually 10X) by the magnification of the objective lens you have pointed at your specimen. For example, a 10X ocular and a 4X low-power objective lens combine for a total magnification of 40X.

Always remember that you are looking at a highly magnified object through a two-lens system. Therefore, the image is always reversed from the way it would appear to your eye. The more an object is magnified, the less depth we are able to see, and focusing becomes more delicate. This is one of the reasons why the fine adjustment focusing knob is used to focus under higher power.

You do maintain some depth perception when using the lower-power objective lenses, but this **depth of field** diminishes to a practical two-dimensional view on all but your lowest power lens. Likewise, your **size of field**, or the area you are viewing is greatest under low power.

It is important to remember that every time you switch to a higher-power objective lens, you are zeroing in on a smaller and smaller part of your specimen; your field of view gets more narrow with each change to higher power. This is why it is important to carefully center the object you wish to enlarge before switching to the next higher objective lens.

ACTIVITY: Learning to Use the Microscope

Even students who have used a microscope in the past may find that they have either forgotten key steps in microscope use, or used vastly different microscopes. Regardless of your experience level, we recommend that you obtain a prepared slide of the letter "e" and follow this step-by-step procedure.

Microscope Use

1. Clip your slide in place on the stage. Center the object to be viewed in the circular hole in the stage. Start with your light turned all the way up and the condenser raised up just below the stage. However, the iris diaphragm should be closed to a pinpoint of light so that your specimen will not look washed out by too much light.

2. Always start with *low power* (the shortest objective lens). This is the only objective lens with which you may use the coarse adjustment knob to find your focus rapidly.

3. Using the two small knobs at the side of your stage, move the slide to scan for an area to enlarge further.

4. Once your specimen is focused and centered, switch to the next higher-power objective lens. *Note:* nearly all microscopes made today are **parfocal**, which means they will automatically be in focus (or close to focused) when you switch to the next objective lens. You should *not* move the adjustment knobs before switching lenses.

5. Use only the fine adjustment knob when you have switched to higher power. *Never* use the coarse adjustment knob with higher power. If you do, you can damage the expensive slide and the much more expensive objective lens.
6. Adjust the iris diaphragm to allow more light through your specimen at higher power.
7. If you lose the focus or placement of your specimen, simply *start over* with Step 1. Do not attempt to carefully move the coarse adjustment. (Starting over is much faster than trying to fix it under high power.) Also, remember to reduce the light through the iris diaphragm when you start over under low power.

Microscope Care

Always carry your microscope with two hands: one hand under the base, and the other holding the microscope's arm.

If the specimen is out of focus or difficult to see no matter how carefully you adjust the focus or the light, then your lens may be dirty. Use the lens cleaner and lens paper provided—do not use paper towels, kimwipes, or tissues. To clean the lenses, dab a drop of lens cleaner solution on the lens paper and rub the ocular and tip of the objective lenses. Then quickly use a dry portion of the piece of lens paper to wipe away the excess solution before it dries; otherwise, it will leave streaks.

When you are finished using the microscope, wrap the cord loosely around your hand and carefully place the wrapped cord over the ocular. Do not wrap the wire around the base where the hot lamp can damage the insulation. Move the nosepiece to the low-power objective and replace the dust cover.

Letter "e" Observation

Obtain a slide of the letter "e" and follow the step-by-step procedure presented in "Microscope Use" above, starting with low power and continuing on to the next higher-power objective lens. Write your observations below.

____

____

____

____

Crossed Threads Observation

Even though the microscopic field of view is essentially two-dimensional, carefully focusing up and down can help your brain visualize the third dimension (depth in this case) to a limited degree. Repeat the above process using the slide of the contrasting color threads. Describe your observations while attempting to focus on these threads.

____

____

____

____

Under which power was it easiest to get both overlapping threads mostly in focus?

____

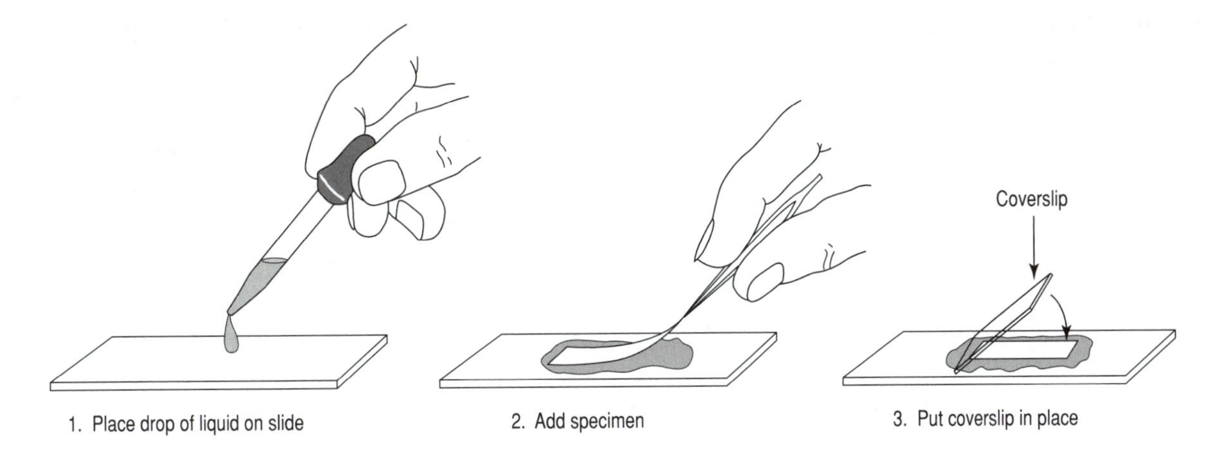

Figure 1.3 Preparation of a wet mount slide.

Preparation of a Wet Mount Slide

Preparation of a wet mount slide is illustrated in Figure 1.3.

1. Obtain a clean blank slide, coverslip, toothpick, and dropper bottles of isotonic saline solution and iodine.
2. Place a small drop of saline solution on the blank slide. *Gently* scrape the inside of your cheek with the toothpick and carefully swirl the scraped cells into the drop of saline solution on the slide. Add one *small* drop of iodine solution to the drop of saline with cells on the slide.
3. Hold the coverslip by the sides so that you do not get fingerprints on it, as you would hold a photograph. Place the edge of the coverslip on the edge of the drop of saline/iodine, lean it toward the liquid, and let go. The coverslip should drop like a door slamming shut on the liquid containing your cells. This technique should minimize the presence of air bubbles under the coverslip.
4. Using the procedure you followed to practice with the letter "e," find your cheek cells first with the low-power objective lens and later with the high-dry lens. *Hint:* Your cheek cells will probably look like magnified, pale-yellow sawdust under low power. Larger objects often mistaken for the cells under low power are usually air bubbles or dirt particles. Unless you have very unusual eyesight, the cells will appear as flakelike cells only after you switch to higher power.
5. Once you have focused the cells with your high-dry objective, adjust your light and compare your findings to Figure 1.4. ■

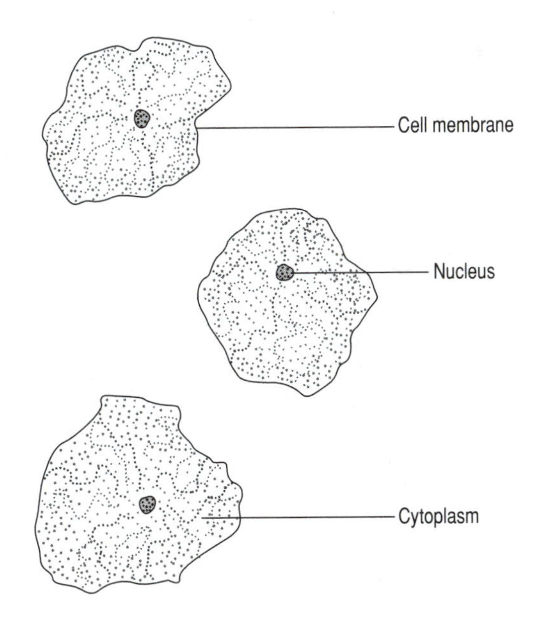

Figure 1.4 Microscopic view of cheek epithelial cells.

Study Questions

1. Summarize the function of the following parts of the microscope.

   a. Nosepiece: ____________________

   b. Iris diaphragm: ____________________

   c. Coarse adjustment: ____________________

2. What does the term **parfocal** mean?

   ____

   ____

3. Why is it important to start with the lowest-power objective lens?

   ____

   ____

   ____

   ____

4. What is the total magnification of your specimen if you are using a 10X ocular and a 45X high-dry objective lens?

   ____

5. Your lab partner accidentally bumps your microscope, and you can no longer see the specimen you were viewing with your high-dry lens. Briefly describe what you should do in order to most safely and efficiently find your specimen.

   ____

6. Why is it so important to center the part of your slide you wish to view when switching to a higher-power objective lens?

   ____

   ____

   ____

   ____

7. Thinking back to the crossed threads slide, what did you learn about depth perception and the microscope?

   ____

   ____

   ____

   ____

   ____

   ____

EXERCISE

Cell Anatomy

OBJECTIVES

After completing this exercise, you should be able to:

1. Identify the major cell structures and their functions.
2. Identify selected cell organelles on models, diagrams, and microscope slides.
3. Understand that cell structure is related to cell function.

MATERIALS

- ❑ Microscope
- ❑ Lens paper
- ❑ Animal cell model
- ❑ Cell anatomy wall chart(s)

Prepared microscope slides:

- ❑ Sperm smear
- ❑ Striated muscle
- ❑ Neuron
- ❑ Ciliated epithelium

Introduction

The cell is the basic structural and functional unit of the body. Depending on their function, cells can differ as to size, shape, and certain internal structures, but all cells possess some common characteristics.

Cells generally have three regions that can be identified with the light microscope: the nucleus, the cell membrane, and the cytoplasm. Other cellular components called organelles are visible only with an electron microscope.

With the light microscope, you will view cell slides and identify the nucleus, cell membrane, and cytoplasm. You will use the cell model, diagrams, and Figure 2.1 to become familiar with the names of the organelles and their locations.

Cell Anatomy

Cell Membrane

The **cell membrane**, or plasma membrane, separates the contents of the cell from the surrounding environment. It is composed of two back-to-back layers of phospholipids in which protein molecules are embedded. Another function of the cell membrane is its ability to be selectively permeable—that is, the cell membrane allows some molecules to pass through it while preventing other molecules from doing so. This selective permeability of the cell membrane controls the movement of substances entering and leaving the cell in both passive and active transport.

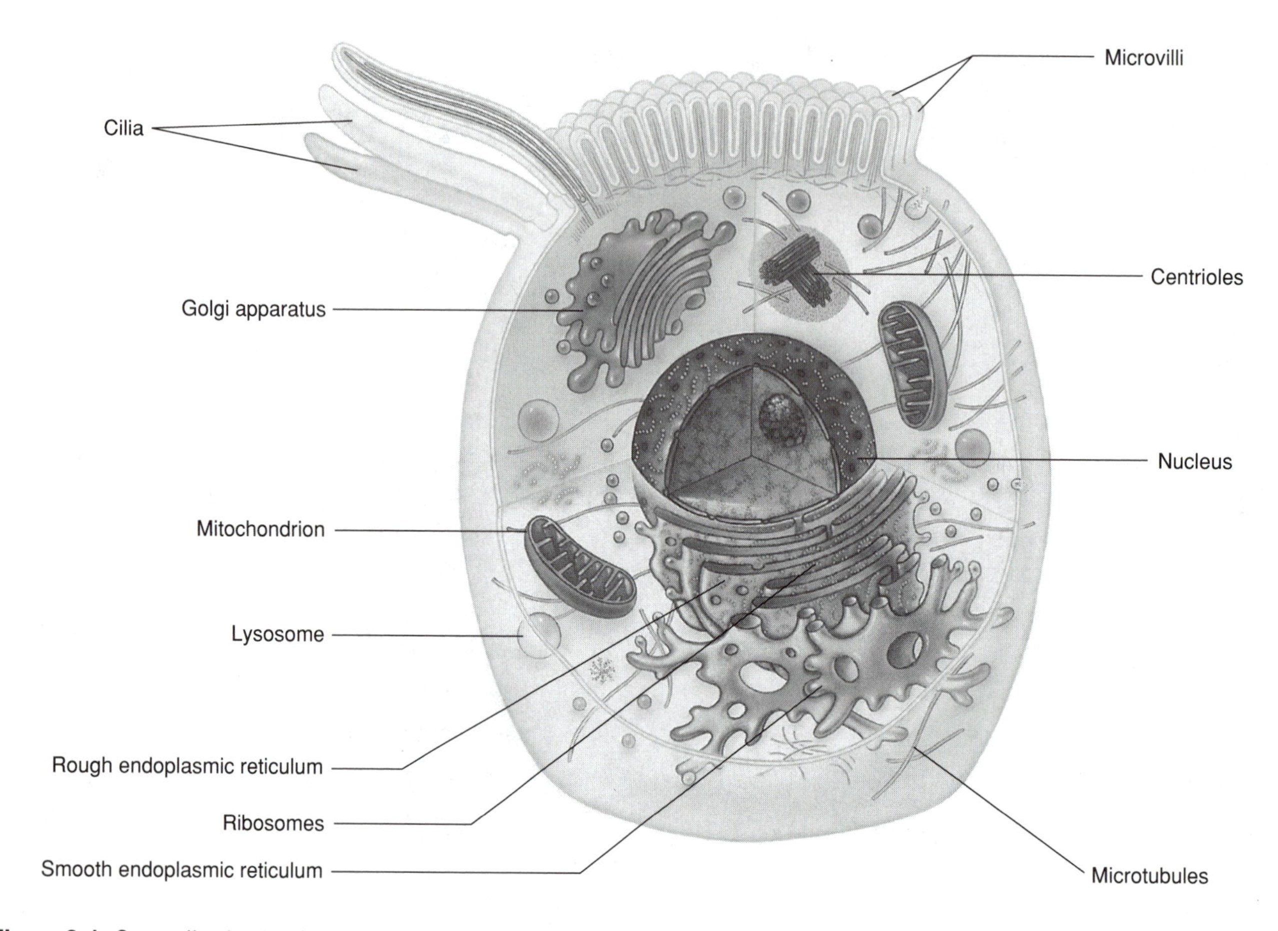

Figure 2.1 Generalized animal cell.

Nucleus

The **nucleus** can be seen with the light microscope as a spherical structure near the center of the cell. It is surrounded by a double membrane called the **nuclear envelope**. The nuclear envelope has pores that allow the movement of material between the nucleus and the cytoplasm.

The nucleus controls cellular activities because it is the site containing genetic material called DNA (deoxyribonucleic acid), threadlike chromosomes that make RNA.

Cytoplasm

The cytoplasm makes up the cell contents outside the nucleus. It is colloid or semifluid. Most activities of the cell are carried out in the cytoplasm by **organelles** (little organs) located there. The organelles are specialized to carry out specific functions for the cell (Table 2.1).

Table 2.1 Summary of Organelles

Organelle	Characteristics	Functions
Ribosomes	Small, round structures. Can be found floating free in the cytoplasm or associated with the endoplasmic reticulum, a complex membrane.	Serve as sites for protein synthesis.
Endoplasmic reticulum (ER)	Network of highly folded membranes. Extends throughout the cytoplasm from the nuclear membrane to the cell membrane.	
	Rough ER—Has ribosomes attached to it.	Provides transport and storage of the proteins formed by the ribosomes.
	Smooth ER—Lacks ribosomes. Is the site of many chemical functions.	Function varies greatly among cells, and includes lipid production, calcium storage, and detoxification.
Golgi apparatus	A stack of flattened sacs located near the nucleus.	Receives proteins from the rough ER and packages them in vesicles for export outside the cell.
Mitochondria	Cylindrical-shaped structures consisting of a double membrane.	Convert energy stored in food to more accessible energy stored in the form of ATP.
Lysosomes	Small membranous sacs containing digestive enzymes.	Contain enzymes that destroy worn-out organelles or foreign particles.
Centrioles	Paired structures composed of microtubules located close to each other and at right angles.	Together with surrounding centrosomes, these function in forming the mitotic spindle during cell division.
Cilia and flagella	**Cilia** are numerous, short, hairlike projections composed of contractile microtubules located on the surface of certain cells.	Used to propel particles across stationary cells.
	Flagella are long, whiplike projections composed of contractile microtubules.	Used to propel a free cell such as sperm.

ACTIVITY: Studying Cell Anatomy

1. Observe the model of a generalized animal cell using your textbook or the key booklet to locate and name the organelles.
2. Using descriptions given from the introduction, wall charts, and the model key booklet, learn and be able to identify structures labeled on the cell diagram (Figure 2.1).
3. Observe Figures 2.2, 2.3, 2.4, and 2.5. Now observe prepared slides of each of these cell types and make a sketch of each as they appear to you under 450X of your microscope. Make sketches in the spaces provided beside each figure. ■

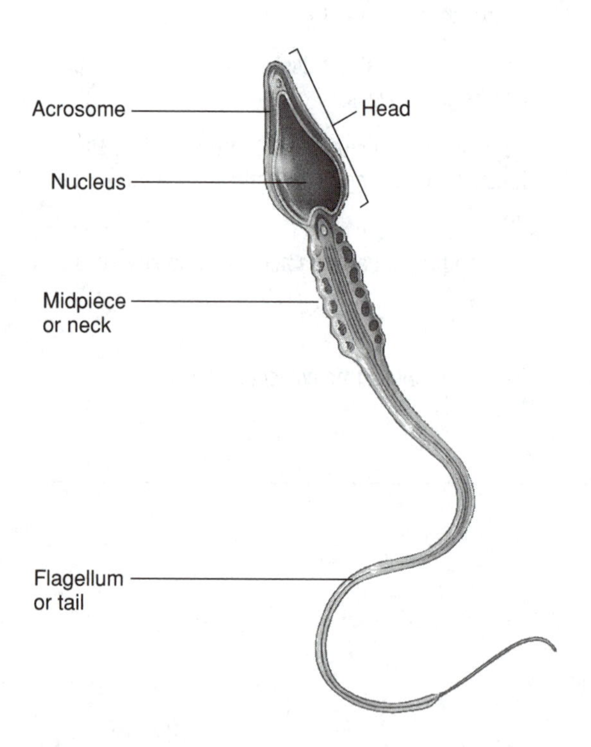

Figure 2.2 Human spermatozoan.

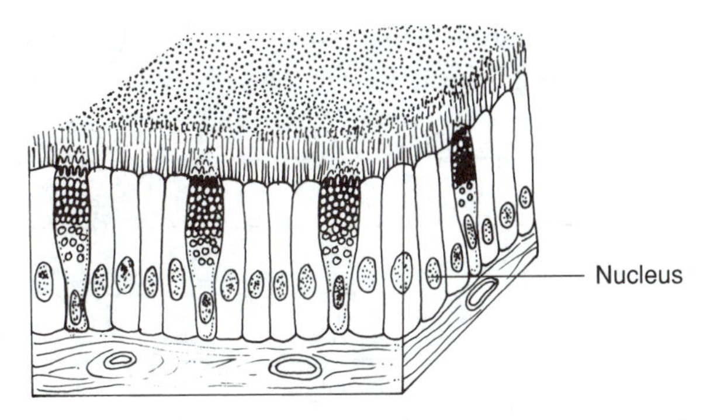

Figure 2.3 Ciliated epithelium.

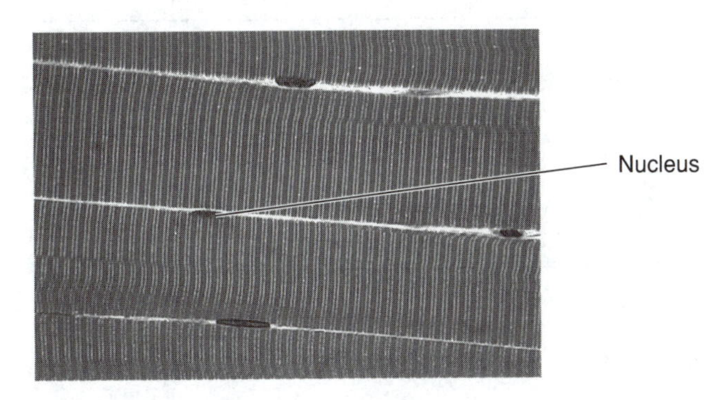

Figure 2.4 Striated (skeletal) muscle cells.

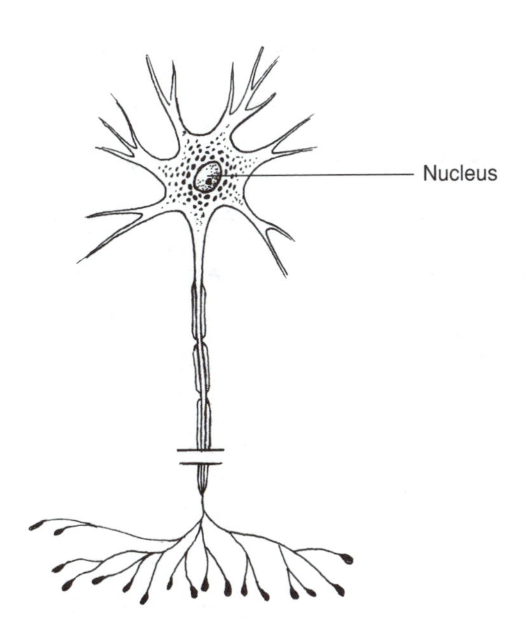

Figure 2.5 Motor neuron.

Study Questions

1. Supply the names of the cell organelles that fit the following descriptions.

   a. "Control box" of the cell; contains DNA and chromosomes

   ____

   b. Cell "suicide sacs" with enzymes

   ____

   c. "Workshop" of the cell; they build proteins

   ____

   d. Two short, rodlike structures that compose the centrosome

   ____

   e. Outer cell boundary with pores

   ____

   f. Tubular network of membranes throughout the cytoplasm

   ____

   g. Potato- or peanut-shaped "powerhouses" of the cell

   ____

   h. Stack of flattened membranes nicknamed "the cell-packaging plant"

   ____

2. Name five cell structures that can be observed with your high-dry objective lens.

   1. ______________________________
   2. ______________________________
   3. ______________________________
   4. ______________________________
   5. ______________________________

3. Name five cell organelles that would be visible only on the electron microscope screen.

   1. ______________________________
   2. ______________________________
   3. ______________________________
   4. ______________________________
   5. ______________________________

4. List three statements in the cell theory.

   1. ______________________________

      ____

      ____

   2. ______________________________

      ____

      ____

   3. ______________________________

      ____

5. What are the functions of cilia and flagella in the human body?

____

____

____

____

____

____

6. List the four cell types you observed and give a brief description of their structure as it relates to their functions.

1. ______________________________

____

____

____

2. ______________________________

____

____

____

3. ______________________________

____

____

____

4. ______________________________

____

____

____

7. Name the three basic parts of a sperm cell.

1. ______________________________

2. ______________________________

3. ______________________________

8. How are striated muscle cells different in appearance from typical animal cells?

____

____

____

____

____

____

EXERCISE

3 Body Terminology, Planes, and Cavities

OBJECTIVES

After completing this exercise, you should be able to:

1. Associate anatomical terms with their meanings as used in textbook and lab manual descriptions and explanations.
2. Identify the various body planes as used in textbook and lab manual drawings and in dissection procedures.
3. Locate the major body cavities and subcavities and identify the major organs located within each cavity.

MATERIALS

- ❑ Human torso model
- ❑ Dissecting probes
- ❑ Wall charts

Introduction

One of the problems a non-science student encounters in a biology course is the use of many Latin or Greek terms. In this course, designed for non-major students, we have tried to keep the use of such terms to a minimum. However, many are part of the language of any text or laboratory manual and, therefore, must be understood by the student. Some of the basic terms that apply to body and organ positions are listed here. Familiarize yourself with them, and they will become part of your language as you progress in the course.

Terminology

Refer to Figure 3.1 and 3.2 when studying the following terms.

- **Anterior** the front end of an animal.
- **Posterior** the rear, or tail end of an animal (also **caudal**).
- **Cranial** the head end of an animal.
- **Dorsal** the back surface of an animal (opposite the belly).
- **Ventral** the belly or undersurface.
- **Lateral** the side, or toward the side.
- **Medial** along the midline.
- **Superior** above.
- **Inferior** below.
- **Superficial** shallow, near the surface.
- **Deep** more internal, away from the surface.
- **Proximal** nearer or closer to the trunk or point of attachment to the body.

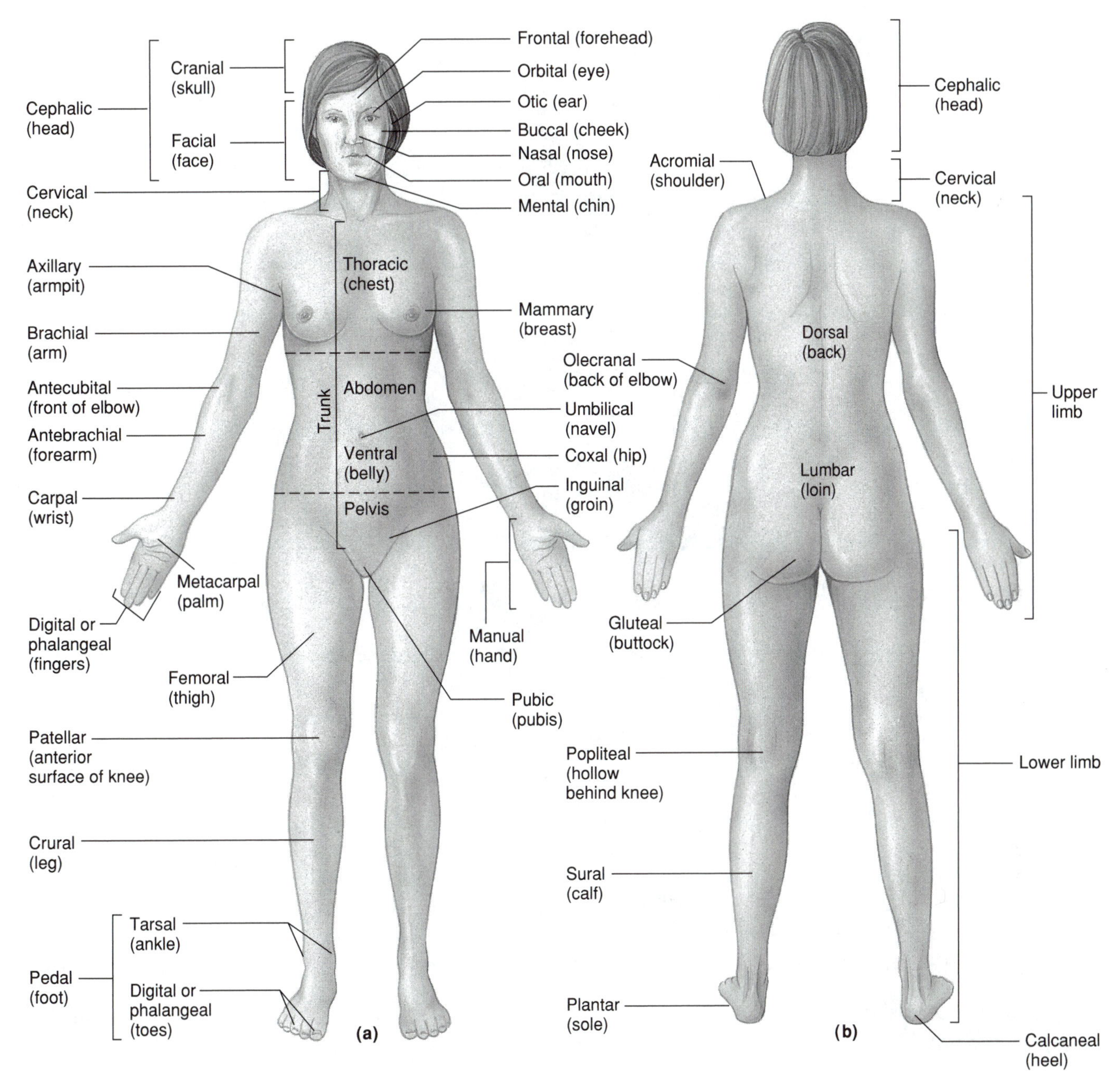

Figure 3.1 Land marks of the human body: a) anterior and b) posterior views.

- **Distal** distant or farther away from the trunk or point of attachment to the body.
- **Prone** lying horizontal on the belly with dorsal side up.
- **Supine** lying horizontal on the back with belly facing up.
- **External** leading to the outside.
- **Internal** leading to the inside.
- **Bilateral symmetry** having two sides or halves alike. This is the most streamlined symmetry (body shape). These animals move faster than animals with other types of symmetry.
- **Anatomical position** standing erect, face view, with arms at sides and palms forward. Notice the body land marks in Figure 3.1.

The rest of this lab session is concerned with body planes and body cavities.

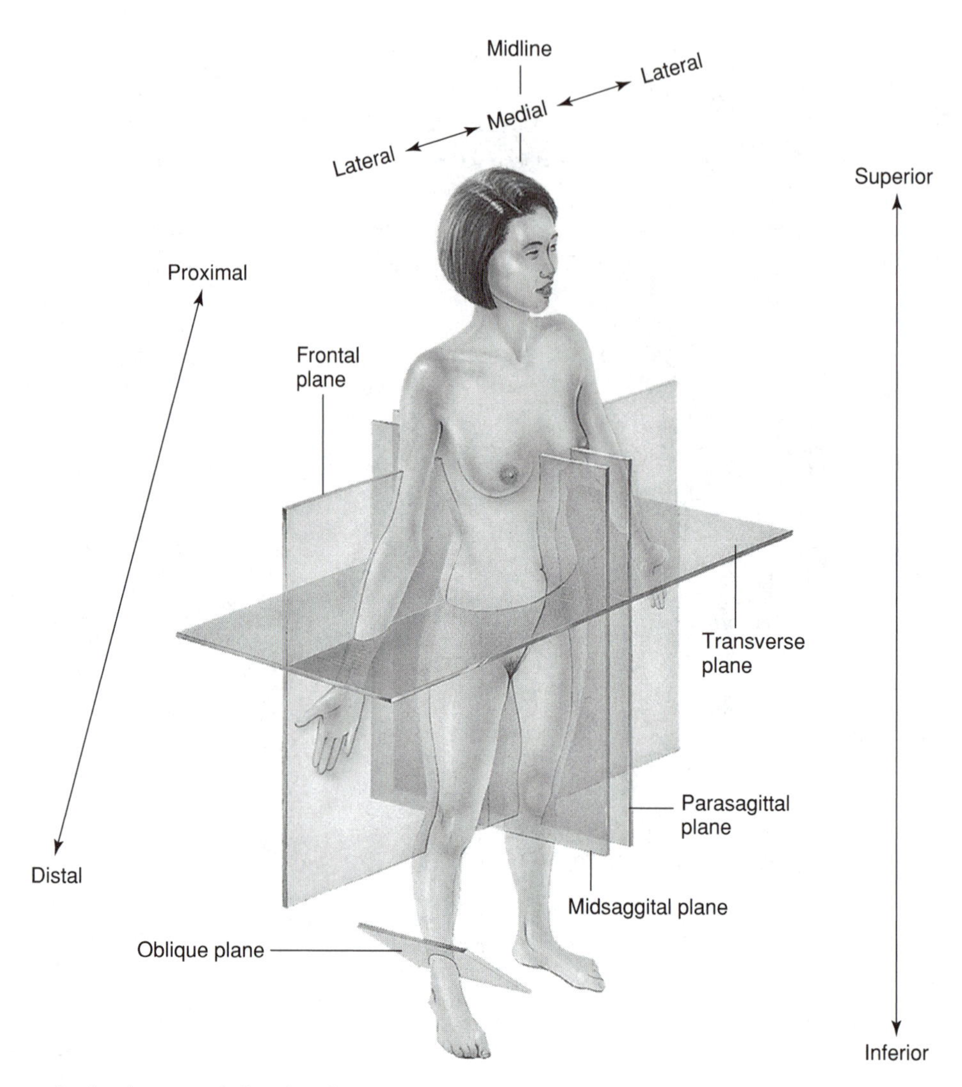

Figure 3.2 Human body planes and directional terms.

Body Planes

Body planes are imaginary or actual dissection cuts through the body that divide the body and its organs into dorsal and ventral, right and left, or other sections. An understanding of the various planes is necessary in understanding dissection procedures and interpreting textbook and laboratory diagrams. Refer to Figure 3.2 as you read and understand the following descriptions of each plane.

- **Frontal plane** a longitudinal plane dividing the body into anterior and posterior halves.
- **Sagittal plane** a longitudinal plane dividing the body into right and left halves.
- **Transverse plane** a "cross cut" through a whole body part, such as the trunk, arm, or leg.

Body Cavities

The body cavities contain various organs (Figure 3.3).

- **Dorsal cavity** in the dorsal region of the body.
- **Cranial cavity** contains the brain.
- **Spinal cavity** contains the spinal cord.
- **Ventral cavity** in the ventral portion of the body.

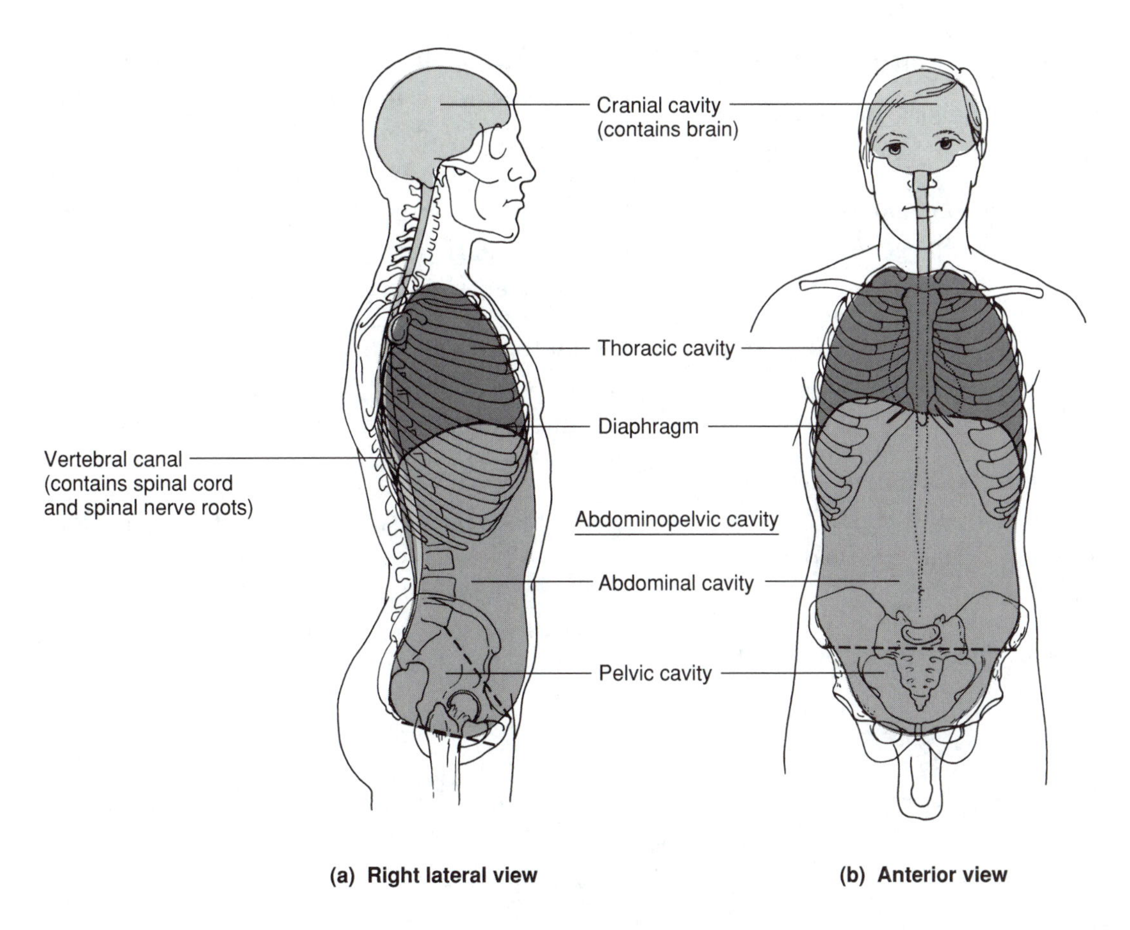

Figure 3.3 Major body cavities.

- **Thoracic cavity** also called the chest cavity; a large cavity located above the muscular diaphragm which is subdivided into:

  Right and left pleural cavities contain the lungs.
- **Mediastium** contains the thymus gland, most of the esophagus, part of the trachea and bronchi, nerves, and lymph vessels, and the aorta and other large blood vessels.
- **Pericardial cavity** contains the heart.
- **Abdominopelvic cavity** a large cavity located beneath the muscular diaphragm. There is no structural separation between the abdominal and pelvic portions.
- **Abdominal cavity** the portion of the general abdominal cavity above the hip bones and below the diaphragm. It contains the stomach, parts of both intestines, and the liver, gallbladder, spleen, pancreas*, and kidneys*.
- **Pelvic cavity** the portion of the general abdominal cavity located within the hip region. It contains the lower parts of the two intestines*, the urinary bladder* and ureters* and, in the human female, the ovaries and uterus*.

*All or part of these organs are retroperitoneal or behind the membrane of the abdominopelvic cavity.

ACTIVITY: Learning Body Terminology

Locate the following on the human torso model and any demonstration specimens chosen by your instructor.

1. Locate the thin, muscular diaphragm, which separates the thoracic cavity from the general abdominal cavity.
2. Look inside the thoracic cavity above the diaphragm and find the two lungs in the right and left pleural cavities.
3. The mediastinum contains the thymus gland, most of the esophagus, and part of the trachea and bronchial tubes, as well as some large blood vessels, nerves, and lymph vessels.
4. Find the four-chambered heart in the pericardial cavity.
5. Now look into the large general abdominopelvic cavity below the diaphragm. In the upper abdominal cavity (from above the hips to the diaphragm), locate the saclike stomach, the upper portion of the large and small intestine, the large reddish-brown liver, the gallbladder, the reddish-brown spleen (on the left side), and the pancreas, and kidneys.
6. Within the pelvic cavity, locate the lower portions of the small and large intestines, the urinary bladder, the ureters, and, if your specimen is a female, the ovaries and uterus.
7. Locate the major organs shown in Figure 3.4. ■

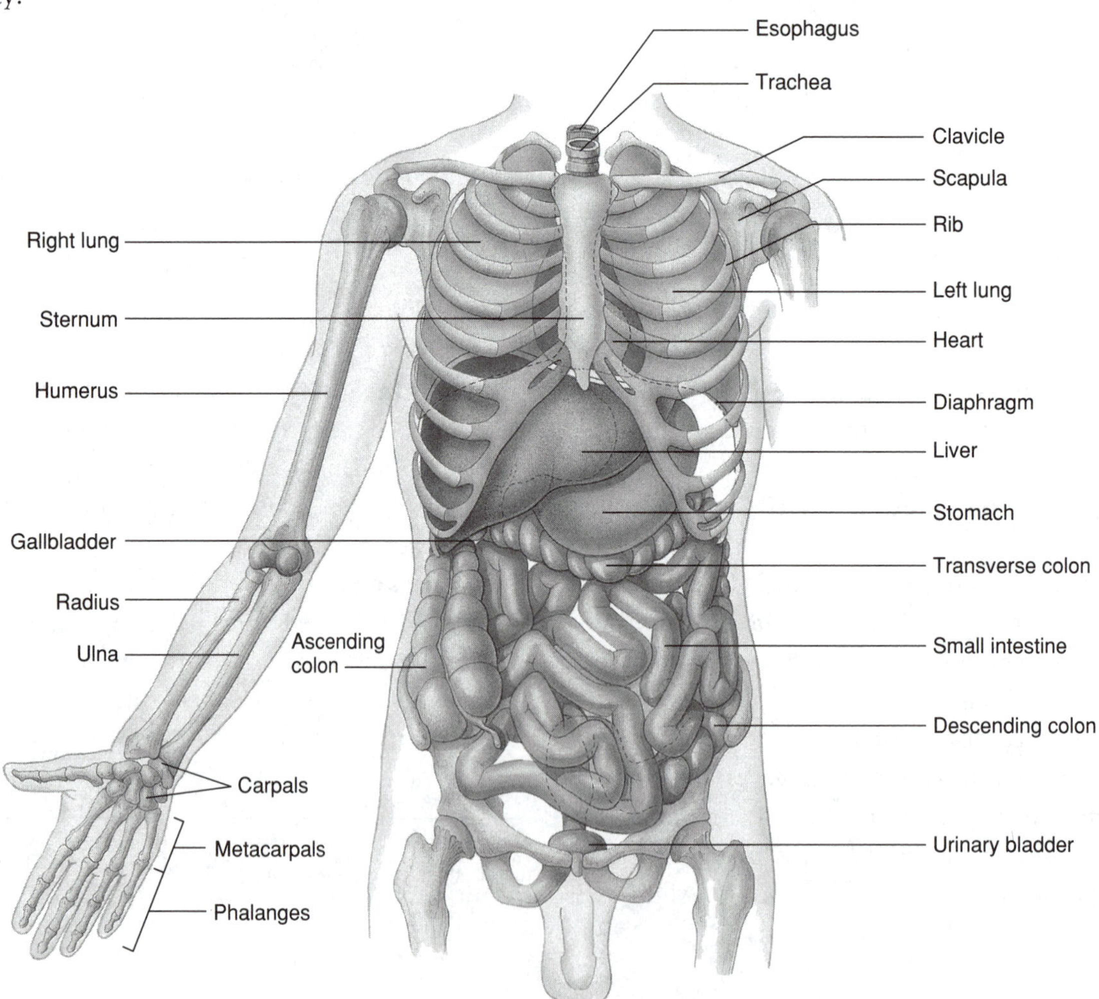

Figure 3.4 Human digestive system and associated structures.

Study Questions

1. Supply the proper term to match each definition.

   a. Lying horizontal on the belly with dorsal side up

   ____

   b. Below or beneath

   ____

   c. Belly or undersurface

   ____

   d. Having two sides alike

   ____

   e. Standing erect, face view, with at least one palm forward

   ____

   f. The front end of an animal

   ____

   g. The head end of an animal

   ____

   h. Leading to the outside

   ____

   i. Near or next to

   ____

   j. The rear or tail end of an animal

   ____

2. Name these body planes.

   a. "Cross cut" through a body part

   ____

   b. Longitudinal plane dividing body into anterior and posterior halves

   ____

   c. Longitudinal plane dividing body into right and left halves

   ____

3. Interpret this lab description: "Make a ventral superficial median incision from anterior to posterior through the skin of your fetal pig."

   ____

   ____

   ____

   ____

4. List five major organs found within the abdominal cavity.

   1. ______________________________
   2. ______________________________
   3. ______________________________
   4. ______________________________
   5. ______________________________

5. List three organs found in the pelvic cavity.

   1. ______________________________
   2. ______________________________
   3. ______________________________

6. Name five organs within the thoracic cavity.

   1. ____________________
   2. ____________________
   3. ____________________
   4. ____________________
   5. ____________________

7. Name the major organ within the:

   a. Cranial cavity ____________________

   b. Spinal cavity ____________________

8. Where is the gallbladder located in:

   a. The fetal pig? ____________________

   b. The human body? ____________________

9. Why is it important to know about the various body terms, planes, and cavities, especially at the beginning of your biology course?

   ____

   ____

   ____

   ____

   ____

10. Describe a median sagittal incision through an organ.

    ____

    ____

    ____

    ____

    ____

11. List four characteristics unique to the mammals.

    1. ____________________
    2. ____________________
    3. ____________________
    4. ____________________

12. Give another name for the mediastinal cavity.

    ____

13. Name three organs within the mediastinal cavity.

    1. ____________________
    2. ____________________
    3. ____________________

EXERCISE

The Tissues 4

OBJECTIVES

After completing this exercise, you should be able to:

1. Define tissue, organ, organ system, and organism.
2. Identify the different epithelial and connective tissues available for study in the lab.
3. Describe the functions of the tissues studied.
4. Describe the locations of the tissues studied.
5. Identify characteristics of the tissues studied.

MATERIALS

Prepared slides of:

- ❑ Simple squamous epithelium
- ❑ Simple cuboidal epithelium
- ❑ Simple columnar epithelium with goblet cells
- ❑ Areolar connective tissue
- ❑ Adipose connective tissue
- ❑ Hyaline cartilage
- ❑ Dense white fibrous connective tissue

❑ Fresh chicken leg

Introduction

The unicellular organism is capable of carrying out all life functions within a single cell, including digestion, excretion, locomotion, and reproduction. With the evolution of multicellularity we begin to see specialization and the development of tissues.

A **tissue** is a group of cells that has become specially adapted to carry out one or several functions. **Histology** is the study of tissues. In the animal kingdom there are four **primary tissue groups,** each specialized for one or several major functional roles in the body (Table 4.1).

With increased complexity, we see that tissues aggregate to form **organs,** organs aggregate to form **organ systems,** and organ systems aggregate to form the **organism.**

Table 4.1 Primary Tissue Groups

Primary Tissues	Major Functions
Epithelium	Covering, lining, protection, secretion, filtration, diffusion, osmosis, absorption
Connective	Binding, connecting, protection, support, storage (i.e., energy, calcium)
Muscle	Contraction, movement
Nervous	Responding to environmental stimuli and transmitting information

Name several organs. What are their functions? Discuss with your class what tissues these organs might be composed of. Place your answers below:

	Organs	Function	Tissues
1.	____________	____________	____________
	____________	____________	____________
2.	____________	____________	____________
	____________	____________	____________
3.	____________	____________	____________
	____________	____________	____________
4.	____________	____________	____________
	____________	____________	____________
5.	____________	____________	____________
	____________	____________	____________

Name several organ systems. What organs are part of these systems, and what is the major function of each system? Place your answers below.

	System	Organs	Function
1.	____________	____________	____________
	____________	____________	____________
2.	____________	____________	____________
	____________	____________	____________
3.	____________	____________	____________
	____________	____________	____________
4.	____________	____________	____________
	____________	____________	____________
5.	____________	____________	____________
	____________	____________	____________

To understand the importance of tissues, it is valuable to review the structural/functional hierarchy of life:

atoms → molecules → macromolecules → organelles → cells → tissues → organs → organ systems → organisms

Thus, tissues provide a critical link between the cellular and organ level.

Epithelial Tissues

Epithelial tissues have several important functions. They may provide protection as they cover and line all cavities, organs, and free surfaces. Secretion of serous fluid, mucous, sweat, hormones, and digestive enzymes are also functions of epithelial tissues. Sometimes they act as a thin barrier, permitting absorption, filtration, diffusion, and osmosis.

Epithelial tissues have the following characteristics:

1. Cells are packed closely together.
2. One surface of the tissue is free, while the other surface is attached to connective tissue below by the non-living basement membrane.
3. The tissues present as one layer of cells (simple) or many layers (stratified).
4. Cells occur in three types:
    a. **Squamous** thin and flat hexagonal cells
    b. **Cuboidal** cubelike cells
    c. **Columnar** rectangular to cylindrical cells (taller than wide)

ACTIVITY: Microscopic Examination of Epithelial Tissue

Each cell type may form tissues through a simple or a stratified arrangement. This lab will focus mainly on the simple tissues. We will see stratified squamous epithelium in the next exercise, *"The Integumentary System."* As you go through the following categories, review the corresponding descriptions and pictures. Use your microscope to view the slides provided.

Simple Squamous Epithelium

Simple squamous epithelium consists of one layer of squamous cells. In surface view, its cells look like floor tiles. When viewed from the side (cross section), the cells appear thin and flat (Figure 4.1).

Locations

The alveoli of the lungs, the glomerulus of the kidney, and the capillaries of the cardiovascular system are major locations of simple squamous epithelium. (Consider what may diffuse through the alveoli, filter through the glomerulus, and diffuse through the capillaries.) Serous membranes lining the ventral body cavity demonstrate the secretory role occasionally given to simple squamous epithelial tissues.

How is simple squamous epithelium adapted to diffusion, filtration, and osmosis?

____

____

____

____

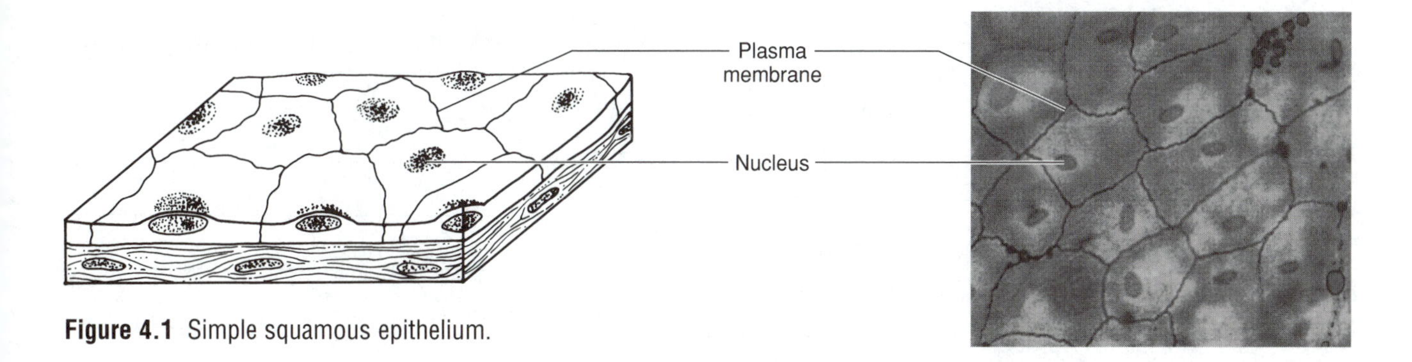

Figure 4.1 Simple squamous epithelium.

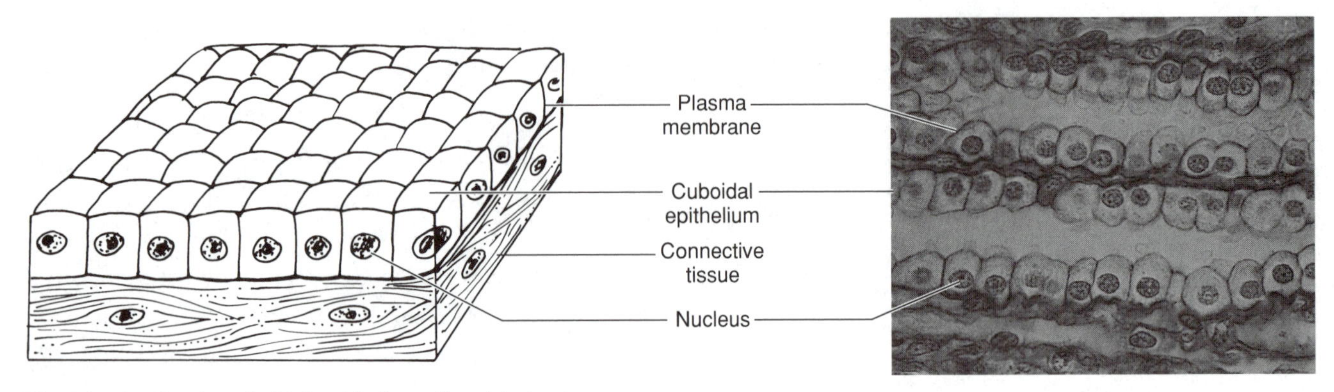

Figure 4.2 Simple cuboidal epithelium: thyroid gland.

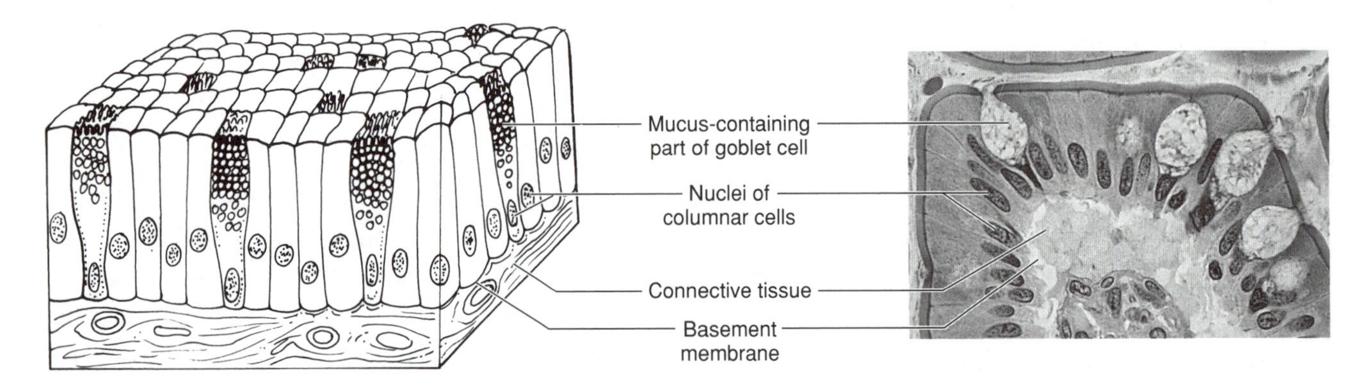

Figure 4.3 Simple columnar epithelium: intestine

Simple Cuboidal Epithelium

Cells of this tissue are cubelike, and of the same height, width, and depth (Figure 4.2).

Locations

Simple cuboidal epithelium lines the ducts of glands and forms ducts and follicles of glandular organs. Although the major function is secretion, cuboidal tissues may also be involved in absorption (i.e., kidney tubules).

Simple Columnar Epithelium

An example of simple columnar epithelium of the intestine is found in Figure 4.3.

Locations

Simple columnar epithelium lines most of the digestive system. These tissues may have microvilli, cilia (in the **pseudostratified** variety), or goblet cells. Columnar epithelium is important for absorption, secretion of enzymes and mucous, and movement of materials (ciliated).

1. How is columnar epithelium adapted to secretion?

____

____

____

____

2. What is the role of cilia in the respiratory system?

____

____

____

____

Fill in Chart 4.1 as a self-test. ■

Chart 4.1

Epithelial Tissue	Sketch	Function	Location
Squamous			
Cuboidal			
Columnar			

Connective Tissues

The connective tissues are very diverse in structure and function. Varied cells scattered in a matrix with varied fibers make some connective tissues look more like building materials than living tissues. Functions of connective tissues include protection, binding/connecting, storage (i.e., of energy, calcium), fighting infection, and support.

Connective tissues have the following characteristics:

1. Varied cells are scattered in a matrix with varied fibers.
2. The nonliving matrix of intercellular material includes gel, solid, and liquid varieties.
3. The numerous fibers of these tissues include collagen, elastic, and reticular fibers.
4. These tissues are formed by numerous types of cells and matrix patterns, including bone cells, fiber-producing cells, adipose cells, cartilage cells, and blood cells.

ACTIVITY: Microscopic Examination of Connective Tissue

As you go through the following categories, review the corresponding descriptions and pictures. View the slides provided using your microscope.

Hyaline Cartilage

An example of hyaline cartilage is found in Figure 4.4.

Locations and Functions

Hyaline cartilage forms a large proportion of the fetal skeleton. It also forms articular cartilage (at movable joints) in the nose (tip), ribs (anterior/medial ends),

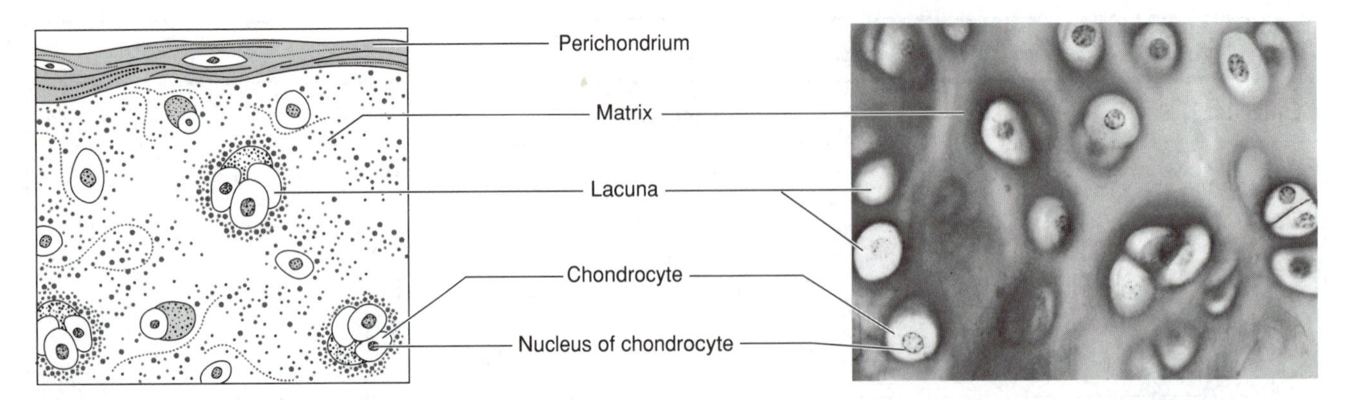

Figure 4.4 Hyaline cartilage.

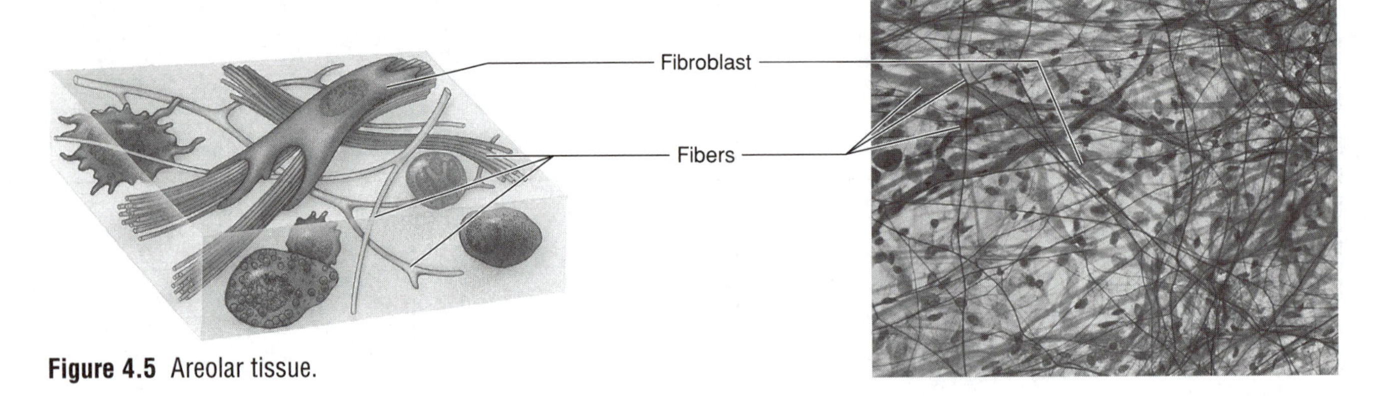

Figure 4.5 Areolar tissue.

and portions of the larynx and trachea. This tissue acts to protect bones and provide flexible linkage between bones.

1. What is the arrangement of cells in this tissue?

____

____

____

____

2. What is the nature of the matrix in this tissue?

____

____

____

____

Areolar Tissue

An example of areolar tissue (loose connective tissue) is found in Figure 4.5.

Locations and Functions

Areolar connective tissue is found under the skin and around nerves, blood vessels, and other organs. As the body's "packing material," its loose, somewhat spongy texture helps it protect delicate organs.

1. Describe the arrangement of cells and fibers in areolar tissue.

____

____

____

____

2. Do all cells and fibers in this tissue look the same?

____

____

3. What type of matrix does this tissue form?

____

____

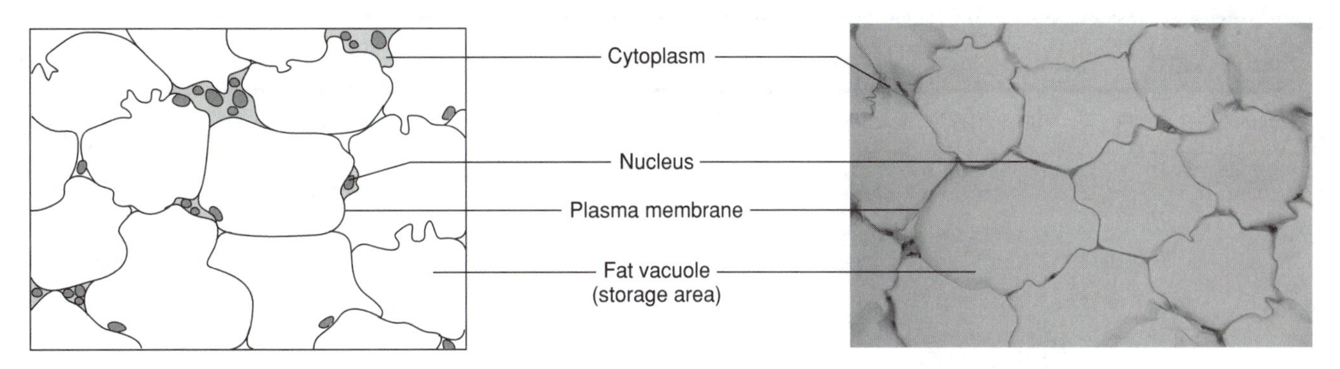

Figure 4.6 Adipose tissue.

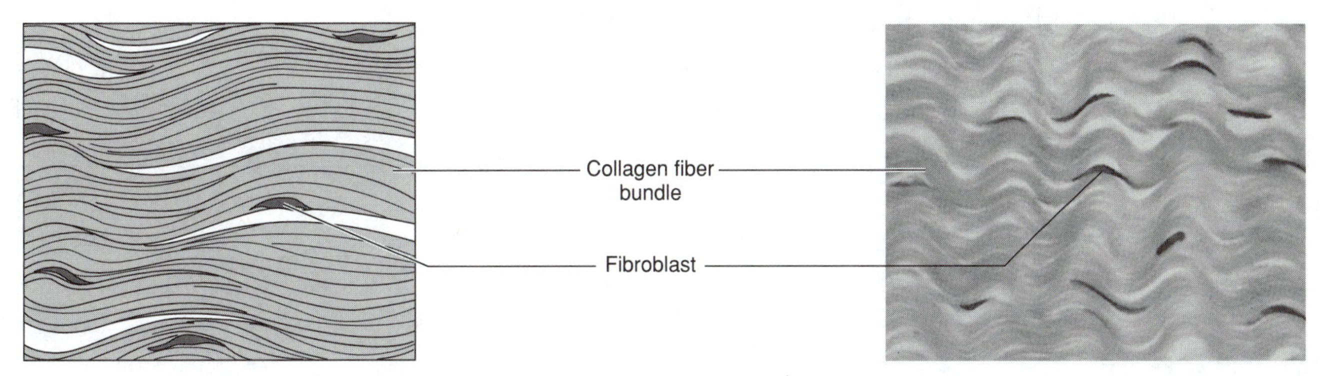

Figure 4.7 Dense white fibrous connective tissue: tendon.

Adipose Tissue

An example of adipose tissue is found in Figure 4.6.

Locations and Functions

Adipose tissue is found beneath the skin, around the kidneys, behind the eyes, and around certain kinds of muscle tissues. This tissue stores energy, acts as insulation, and provides protection.

1. Does this tissue have a matrix?

____

____

2. This tissue provides a function like one of the other tissues: what function and which tissue?

____

____

3. Where is fat stored in this tissue?

____

____

Dense White Fibrous Connective Tissue

An example of dense white fibrous connective tissue is shown in Figure 4.7.

Locations and Functions

This tissue is found in tendons, ligaments, and deep fascia. Its main feature is its strength. It resists tearing due to the large number of collagen fibers; thus, its role is to bind, protect, and connect.

1. What is the arrangement of fibers in this tissue?

____

____

____

____

2. Can you see this tissue's matrix?

____

____

Chart 4.2

Tissue	Sketch	Function	Location
Epithelial			
Hyaline cartilage			
Areolar			
Adipose			
Dense white fibrous			

Note: There are numerous other connective tissues, such as bone and blood, and your instructor may choose to present these tissues during this lab. We will study bone in Exercise 6, Skeletal System, and blood in Exercise 10, Circulatory System. Muscle and nervous tissues will be covered in the chapters about those systems. ■

Study Questions

1. The chicken leg is an example of an organ. Which tissues can you identify in this organ?

___

___

___

___

2. Fill in Chart 4.2 as a self-test.

EXERCISE 5

The Integumentary System

OBJECTIVES

After completing this exercise, you should be able to:

1. Identify the major components of skin.
2. Name the tissues found in skin.
3. Identify the epidermis and dermis on models and slides.
4. Enumerate the functions of skin.

MATERIALS

- ❑ Model of skin and the key to the model
- ❑ Slides of skin

Introduction

While the skin is widely considered the simplest organ system, it is also the largest in terms of surface area, and its functions are essential for human survival. It is considered a system because it has elements of all four tissue groups and has organs such as nails, hair, sweat glands, and oil glands.

Functions

The integumentary system provides a variety of services for the human body. It protects us from bacteria, ultraviolet (UV) radiation, and dehydration. As a major sensory system, it includes receptors for hot, cold, touch, pressure, and pain. The skin plays a crucial role as our main thermoregulatory apparatus, regulating body temperature by vasodilation, vasoconstriction, and sweating. It synthesizes vitamin D in the presence of sunlight. Finally, the skin is a minor excretory organ, eliminating water, salt, and urea.

Structure

The skin (Figure 5.1) is composed of two regions, the upper **epidermis** and the lower **dermis**.

Epidermis

The epidermis is the region of the skin composed mainly of **stratified squamous epithelium**. You may recall from the tissue exercise that stratified squamous epithelium is a tissue composed of several layers of flat cells. As is true of most epithelial tissues, the epidermis has *no* blood vessels of its own, and the blood supply on which it depends for survival is found in the dermis below.

There are five layers of cells in the stratified squamous epithelium of the epidermis. The shape of the

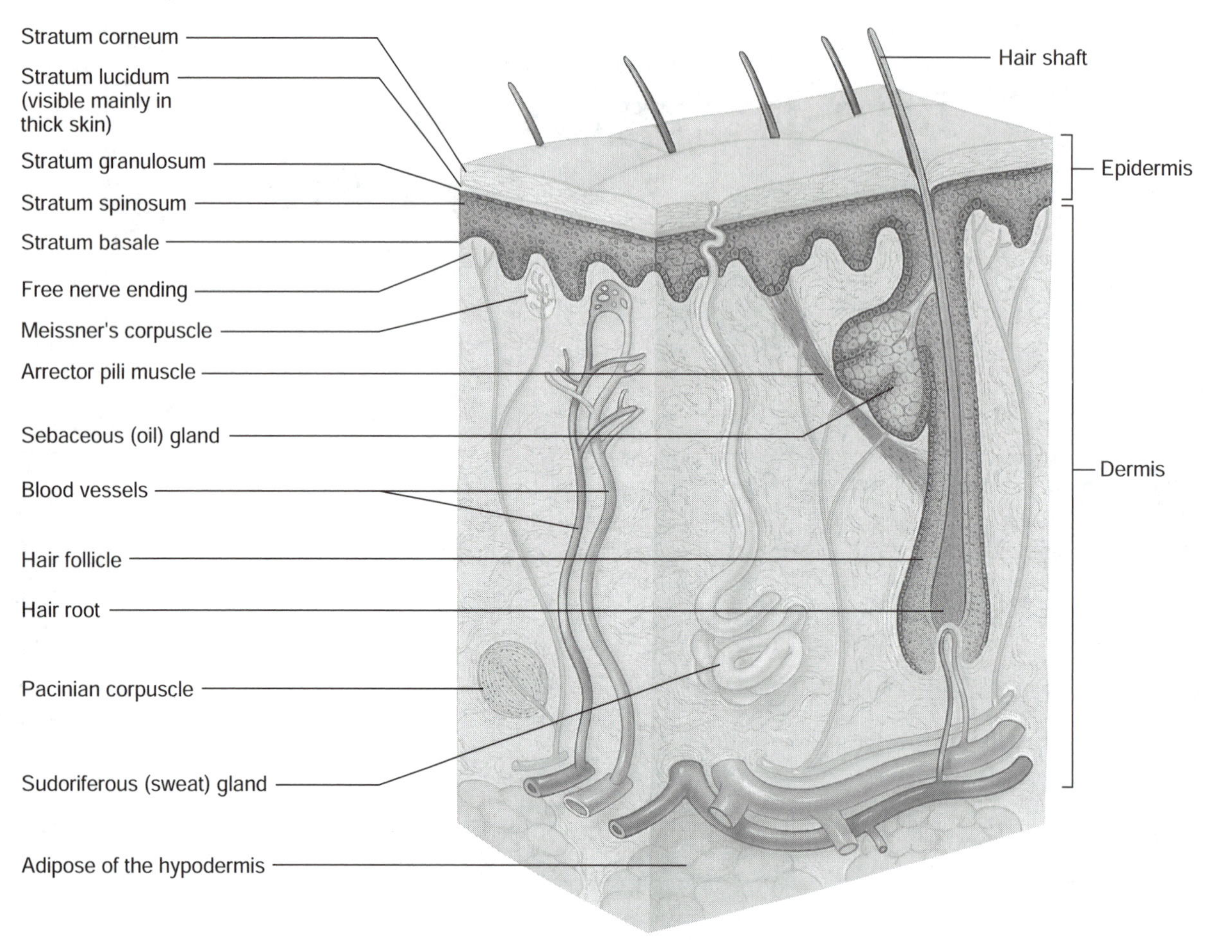

Figure 5.1 Cross-section of skin.

cells changes and the cells die as they are pushed toward the surface, away from the dermis and its blood vessels carrying nutrients and oxygen.

Although most epithelial tissues do not have anything other than their particular variety of cells, the epidermis has some **free nerve endings** that function as pain receptors and pigment synthesizing cells.

Epidermal Layers

The **stratum basale** is the bottom or deepest layer of cells, where the highly mitotic cells appear cuboidal in shape. This layer is attached to the **basement membrane**. These are the only cells that undergo nearly constant **mitosis,** replacing old dead cells shed from the top layer. Also found here are **melanocytes**, scattered throughout the row of cells. Melanocytes synthesize the brownish-black pigment **melanin**, which protects the skin from excess UV (from sunlight) injury.

The **stratum spinosum, granulosum,** and **lucidum** have many rows of polyhedral cells that slowly die as they are pushed farther away from their food supply in the dermis. They begin to undergo the complex chemical process that synthesizes keratin. The farther they get from the stratum basale, the flatter and thinner these cells become. The waxy protein that waterproofs skin, keratin, is not present in these three layers, but is being synthesized.

The uppermost layer of skin, the **stratum corneum,** is composed of many rows of squamous cells. The thickness of this layer depends on exposure to friction (thickest on the soles of the feet, thinnest at the eyelids and lips). These flat cells are dead and filled with keratin, and are shed constantly.

Dermis

The dermis is the thicker and more complex lower region of skin, mainly composed of a tissue similar to areolar connective tissue, but possessing a greater variety of fibers in a unique configuration. This mix of reticular, collagen, and elastic fibers makes the dermis strong, elastic, and a good tissue bed for the many structures found here. It is the arrangement and folding of the dermis that forces the epidermis to fold (this is seen as fingerprints).

Structures of the Dermis

Blood vessels are located deep in the dermis or in the hypodermis ("below the dermis"). They form capillary loops.

Nerve receptors (refer to Figure 5.1 for appearance) include *Meissner's corpuscles* and *Pacinian corpuscles*. **Meissner's corpuscles,** specialized nerve endings located in the dermal papilla, are sensitive to light touch. **Pacinian corpuscles**, large, specialized nerve endings located deep in the dermis and hypodermis, are sensitive to pressure.

Hair follicles are portions of the stratum basale of the epidermis that grow down into the dermis, forming a tubelike channel. At the base of the hair follicle is a pore through which nerves and blood vessels enter the follicle. A small mass of mitotic tissue divides to form the cells of the hair, which are pushed up and out of the follicle. The **hair root** is the portion of the hair in the follicle, while the **shaft** is the portion of the hair that projects out of the follicle. The **cuticle** is the waxy, clear protective outer layer and the **medulla** is the hollow innermost portion. The **cortex** is the pigmented portion of the hair.

Arrector pili muscles are small, smooth, involuntary muscles running from the hair follicle to the stratum basale. Contractions of this muscle pull the hair into an erect position. The rationale behind this is simple—erect hair forms a thicker layer of insulation in mammals. The skin is pulled upward, producing "goose bumps" in humans (we lack enough hair to obtain any significant insulation benefits).

As the stratum basale folds inward to form the hair follicle, **Sebaceous glands** are also formed. These saclike glands secrete a oily substance called **sebum**. Sebum coats the hair and prevents the dryness that would otherwise make the hair brittle and vulnerable to breakage. If the follicle becomes obstructed, bacteria trapped in the follicle digest sebum, forming fatty acids that irritate the follicle and cause inflammation. White blood cells migrate in, forming a pimple or whitehead.

Sudoriferous sweat glands are coiled tubular glands which are formed from stratum basale folding down into the dermis. They secrete a watery substance containing some dissolved salts, called sweat. **Eccrine sweat glands** are found over most of the body, and secrete a watery, slightly salty solution. They are essential for thermoregulation as evaporation dissipates heat. **Apocrine sweat glands** are large coiled sweat glands located in the groin and axilla. Its secretions contain lipid and protein, which are fermented by bacteria, producing odorous end products.

Nails are formed in a similar manner to hair. The stratum basale folds down into the dermis, pushing up the keratin-producing cells that form the nail. Like hairs, nails are living at the base, but die as they are pushed up and away from blood vessels carrying nutrients and oxygen. Dead nail cells become keratinized.

Hypodermis

Below the dermis is the hypodermis, also called subcutaneous tissue or superficial fascia. There are conflicting schools of thought on the issue of whether the hypodermis is part of the integumentary system. It is usually not considered part of the skin. It is composed of adipose and areolar connective tissue, has a large number of blood vessels, and is a good site for injections. The high density of adipose tissue provides insulation and energy storage.

ACTIVITY: Microscopic Examination of the Integumentary System

1. Observe the model of skin and identify all of the structures mentioned in the discussion.

2. Observe the slide of skin. Try to position the slide so the dermis is down and the epidermis is up. Identify the structures described in the discussion. Do they differ from the model, photo, and drawing?

____

____

3. Sketch the skin in the space provided below.

4. Based on the slide, indicate how it differs from the model and Figure 5.1.

____

____

____

5. Why do the structures look different from the model? (*Hint:* Remember this is a section.)

____

____

____ ■

EXERCISE

The Skeletal System

OBJECTIVES

After completing this exercise, you should be able to:

1. Understand major functions of the human skeleton and its relationship to other systems.
2. Locate and name major bones of the body.
3. Understand various joint types of the body.
4. Note differences between the human male and female skeleton.
5. Understand the internal structure of long bones.
6. Note differences between the fetal and adult skull.

MATERIALS

- ❑ Human skeleton
- ❑ Plastomount of human ear bones
- ❑ Male and female pelves
- ❑ Fresh long bone, cut in long section
- ❑ Fetal skull
- ❑ Prepared slides of compact bone showing cross-section of Haversian systems
- ❑ Charts of bone, cartilage, and the human skeleton

Introduction

Humans and other vertebrates possess a jointed **endoskeleton,** or a skeleton inside the body. The adult human skeleton consists of 206 separate bones, which are fully ossified (formed of bone) by about age 21. Bones of the human fetus and, to a lesser extent, the newborn child are mostly cartilage and more solid than adult bones. They elongate, hollow out, and ossify during the growing years.

The skeleton provides support, protects internal organs, produces blood cells, stores minerals (calcium and phosphate) and energy (yellow fat), and provides levers for locomotion.

Osseous Tissues

All bones contain two types of osseous tissue—a soft **spongy** type and a solid **compact** type (Figure 6.1).

Spongy bone is formed from spicules of bone deposited in an irregular pattern. It is lighter and more fragile than compact bone, and contains blood-forming tissue. In adults, spongy bone is the main site of **hemopoiesis** (blood cell production). Spongy bone makes up the interior of cranial bones and the epiphysis of long bones.

Compact bone is dense, solid bone deposited in a **Haversian system**, a highly regular pattern that forms around a blood vessel. Compact bone is found on the exterior surface of most bones, and makes up

(text continues on page 35)

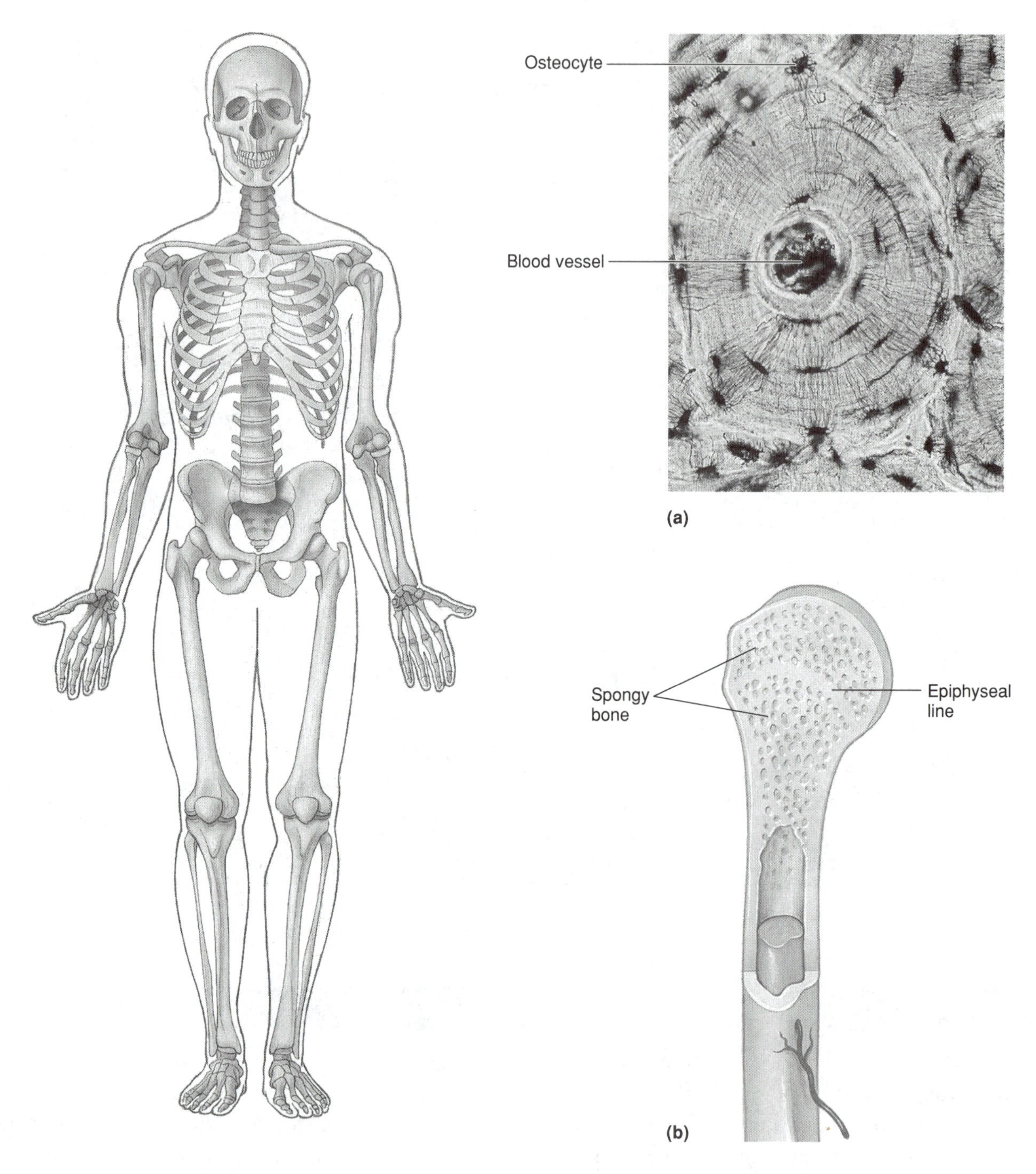

Figure 6.1 The bony framework of the skeleton consists of (a) compact bone and (b) spongy bone.

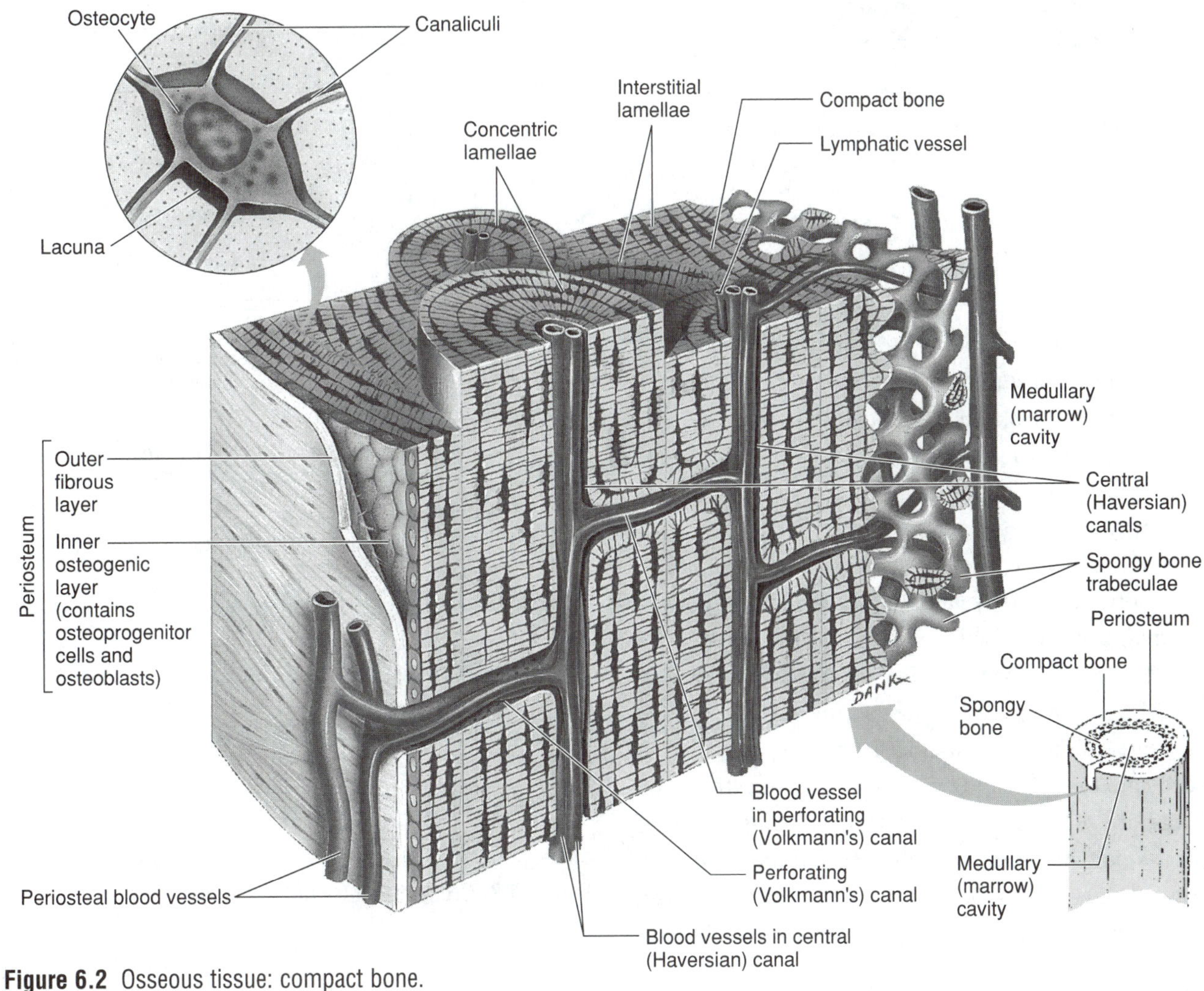

Figure 6.2 Osseous tissue: compact bone.

the entire bony tissue of the diaphysis (shaft) of long bones.

A **Haversian system** or **osteon** is a pattern of bone formation in which the osseous tissue is deposited in rings due to the circular arrangement of bone cells around a blood vessel. Parts of a Haversian system include:

- **Haversian (central) canal** the microscopic central cavity in a Haversian system in which the blood vessel and nerves are located.
- **Lacunae** rings of small spaces in the osseous tissue, filled with the living bone cells.
- **Osteocyte** the bone cells found in the lacunae.
- **Lamellae** rings of osseous tissue.
- **Canaliculi** small canals in the osseous tissue; provide a path for nutrients to travel to the osteocytes.

ACTIVITY: Microscopic Examination of Compact Bone

1. Obtain a prepared slide of osseous tissue.
2. Referring to Figure 6.2, examine the tissue and identify the structures indicated. ■

Bone Classification

Bones are often classified according to shape. The most complex of these are **long bones**. Following is a description of the parts of a long bone (Figure 6.3).

Gross Anatomy of a Long Bone

The **epiphyses** are knoblike ends of the long bone. They are composed of spongy bone covered with a thin layer of compact bone. Hemopoiesis takes place in the spongy bone tissue of the epiphyses. The **epiphyseal plate**, a plate of cartilage, is a place where rapid growth in length occurs during childhood. When growth in length ends, it leaves an ossified remnant called the **epiphyseal line**.

The **diaphysis** is the hollow shaft of a long bone. The hollow **medullary cavity** is a site of hemopoiesis in a child, but it fills with yellow fat in the adult.

On the outside of the bone, the **periosteum** is a fibrous membrane that protects the surface of the long bone. Blood vessels arise from the periosteum. At the epiphyses, a cap of cartilage covers each end of a long bone. This is called **articular cartilage**, as it is typically part of a movable joint. The inner surface of the medullary cavity is lined with the **endosteum**, which is composed of phagocytic osteoclasts.

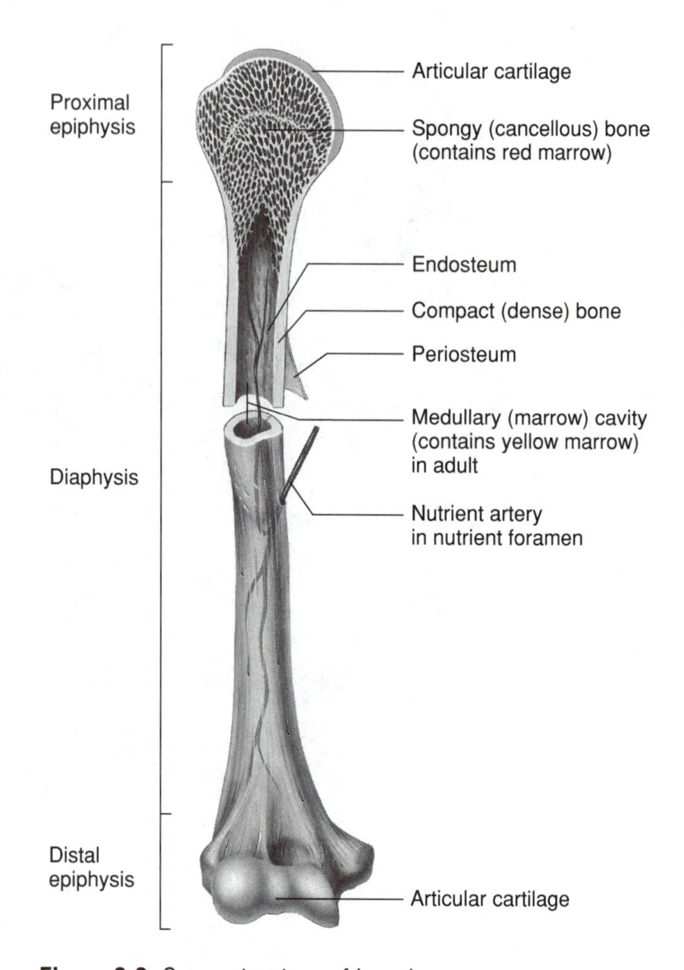

Figure 6.3 Gross structure of long bone.

ACTIVITY: Examining a Long Bone

Examine a longitudinally sectioned long bone and locate the parts identified in Figure 6.3. ■

The Infant Skeleton

The infant skeleton is quite different from that of the adult. The head is proportionately large in relation to the rest of the body, the chest is more rounded in contrast to the oval-shaped chest of the adult, and the face is small in proportion to the cranium. Since many of the bones have not finished growing, many are still separated by softer connective tissues until they fuse later in life. Many soft spots, or spaces spanned by connective tissue between bones called **fontanels**, are present in the infant skull (Figure 6.4).

ACTIVITY: Observing the Infant Skull

1. Observe the infant skull on demonstration. *Do not handle it, as it is very fragile.*
2. Notice the fontanels, the connective tissue between bones where bone formation is not yet complete.
3. Locate the **anterior, sphenoid, posterior, and mastoid fontanels.** See Figure 6.4. ■

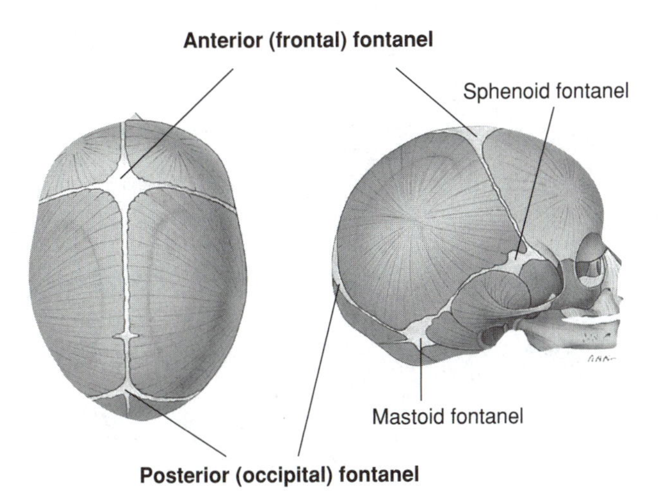

Figure 6.4 Infant skull.

The Adult Skeleton

The human skeleton (Figure 6.5) consists of two main divisions: the **axial division** and the **appendicular division**.

The Axial Skeleton

The axial division runs along the long (longitudinal) axis of the body and is part of the skeleton without the appendages. It includes the skull, 3 bones of each middle ear, 33 vertebrae, the hyoid bone, 12 pairs of ribs, and the flattened sternum or breastbone.

ACTIVITY: Observing the Axial Skeleton

1. Observe the adult human skeleton.
2. Locate all of the bones and structures indicated.
3. Compare the adult and fetal skull.

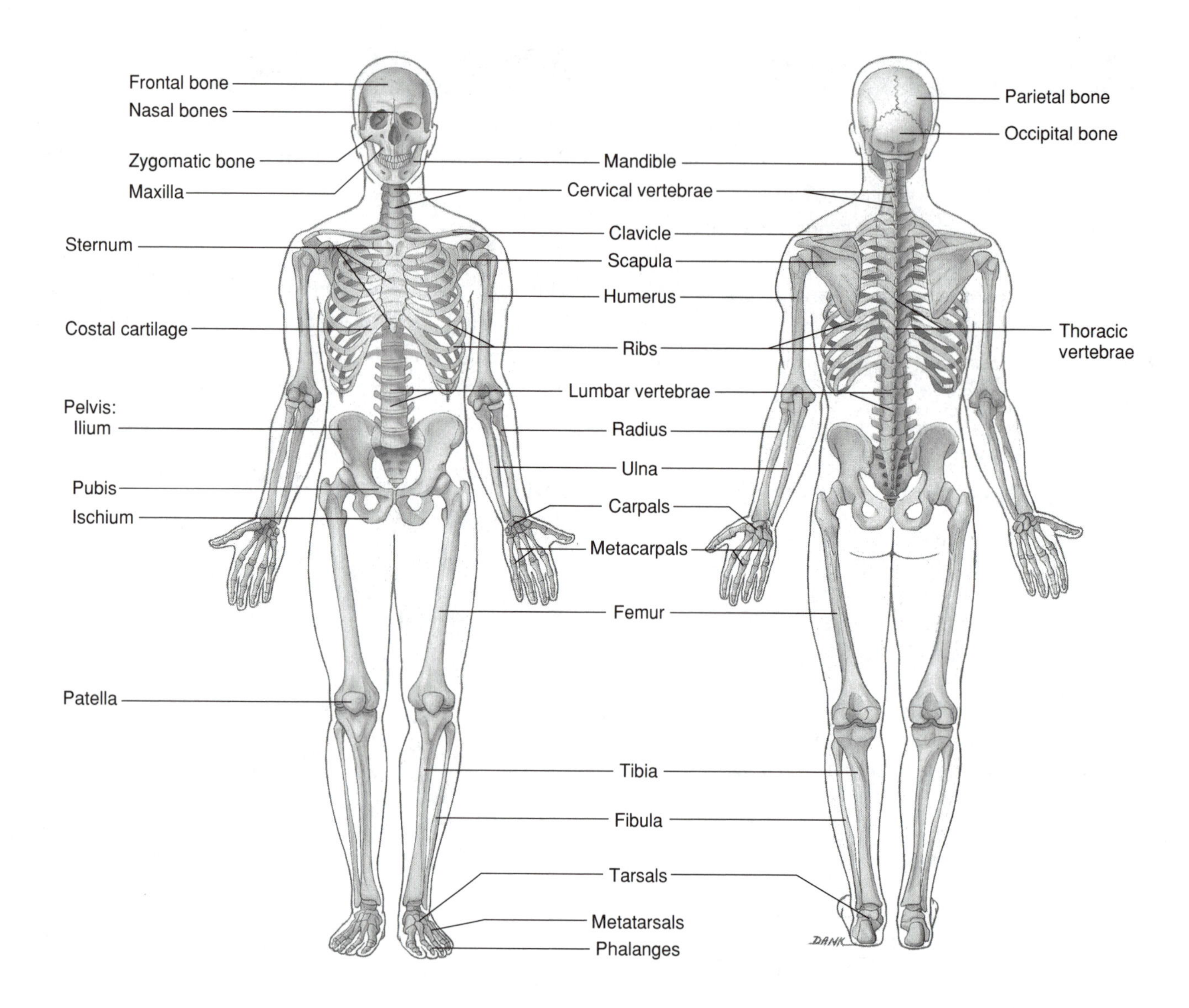

Figure 6.5 Skeletal system.

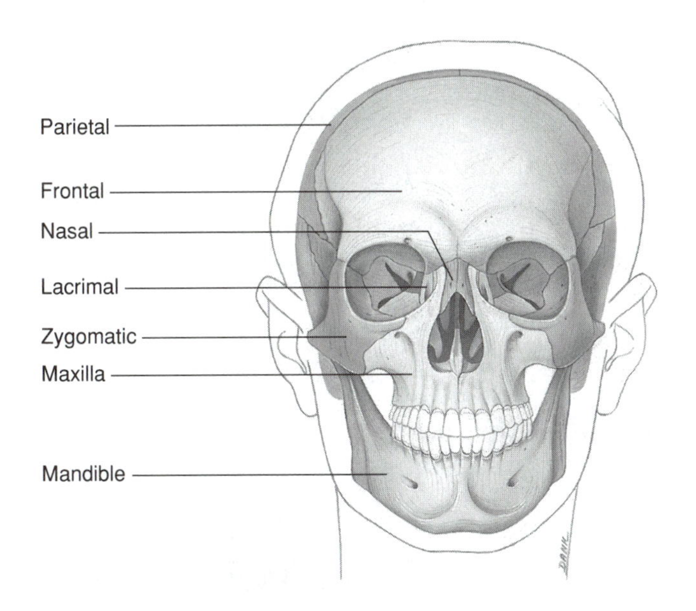

Figure 6.6a Human skull: anterior view.

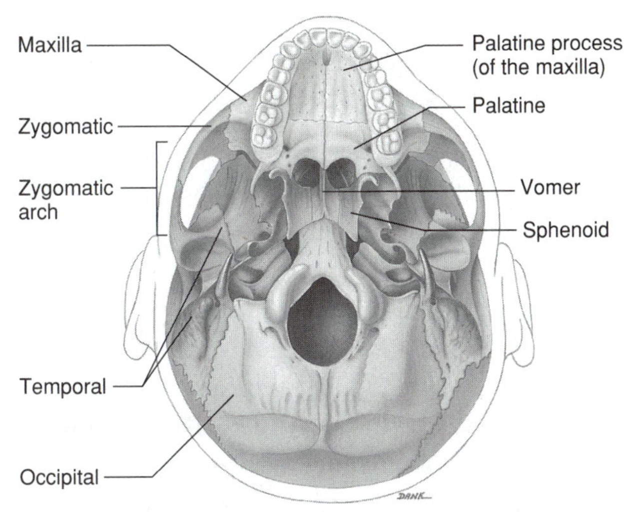

Figure 6.6c Human skull: inferior view.

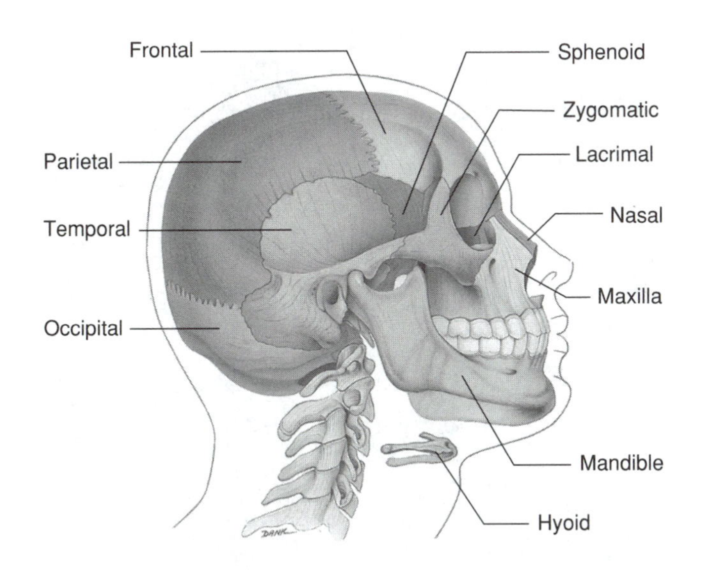

Figure 6.6b Human skull: left lateral view.

The Skull

The skull includes the bones of the **face,** the **cranium** (brain case), and **jaws.** The upper immovable jaw is called the **maxilla** and the lower movable jaw is called the **mandible.** Both jaws contain long socket-type joints that accommodate **teeth.**

1. Locate the **occipital condyles**, two rounded projections at the base of the skull that pivot on the 1st cervical vertebra as when nodding the head "yes."
2. Observe the bones of the human skull in Figure 6.6a–c. Some of the skull bone names also apply to lobes of the brain (frontal, parietal, occipital, temporal). Find the bones illustrated on a skeleton.

Bones of the Ear

Observe the demonstration plastomount containing the three tiny ear bones, or ossicles. These three bones are present in each middle ear and are moved when air vibrations bend the eardrum inward. More of their function will be studied in a later lab. These three bones have Latin names describing their shapes. They are the **malleus** (hammer), **incus** (anvil), and **stapes** (stirrup).

Vertebrae

There are 33 vertebrae, but some of the lower vertebrae are fused (Figure 6.7). These comprise the **vertebral column**, or backbone.

1. Observe the vertebrae listed below:
   - **Cervical vertebrae** the seven neck vertebrae. The first cervical vertebra, or the **atlas**, allows the occipital condyles to slide forward and backward as when nodding the head "yes." The second cervical is called the **axis** and pivots on the atlas as when shaking the head "no." The other five cervical vertebrae are slightly movable against each other in other neck movements.
   - **Thoracic vertebrae** the twelve chest vertebrae. The 12 pairs of ribs arise from these vertebrae posteriorly.
   - **Lumbar vertebrae** the next five vertebrae in the lower back region.
   - **Sacral vertebrae** five fused vertebrae comprising a large triangular-shaped bone between the hips called the **sacrum**.
   - **Coccygeal vertebrae** vestigial human tailbones, usually four, that are collectively called the **coccyx**.
2. Observe the cartilaginous intervertebral discs between each of the vertebrae.

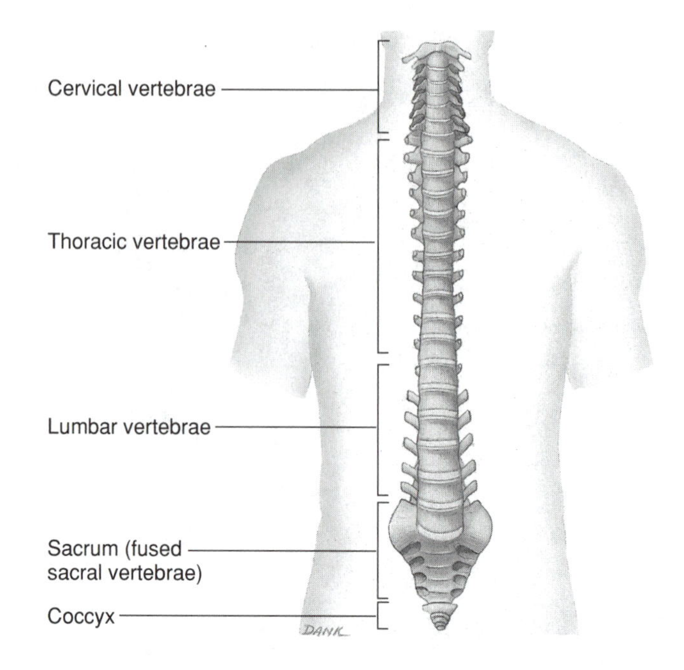

Figure 6.7 Human vertebral column.

Ribs

Twelve pairs of these flattened bones arise from the thoracic vertebrae, and most connect either directly or indirectly to the sternum (breastbone). 12 pairs of ribs are present in both male and female humans (Figure 6.8).

1. Observe the ribs listed below.
   - **"True" ribs** the first seven pairs of ribs, which are connected to the sternum by a short piece of hyaline cartilage called **costal cartilage**.
   - **"False" ribs** the next five pairs of ribs, three of which are indirectly connected to the sternum via the previous rib's costal cartilage.
   - **"Floating" ribs** the last two pairs of ribs, which are not connected to the sternum at all.
   - **Sternum** the flattened breastbone. ■

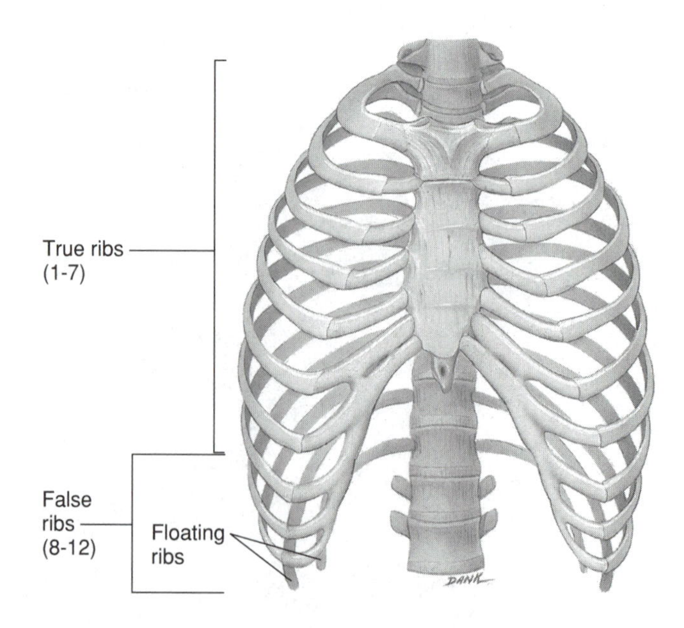

Figure 6.8 Human rib cage: anterior view.

The Appendicular Skeleton

The **appendicular division**, as the name implies, includes appendages—shoulders, arms, hips, and legs. (More detail in this area will be covered in the lab activities that follow.)

Various types of joints between bones allow for numerous body movements and types of flexibility. The lab activities will also cover joint types in more detail, as well as a comparison of the male vs. the female skeleton.

ACTIVITY: Observing the Appendicular Skeleton

1. Observe the adult human skeleton (See Figure 6.5).
2. Locate all of the bones and structures indicated.
3. Compare the male and female pelves.

Shoulder and Arm

Observe the shoulder and arm bones listed below.

- **Clavicle** the collarbone.
- **Scapula** the shoulder blade.
- **Humerus** the single large upper arm bone.
- **Radius** the lateral lower arm bone when in anatomic position.
- **Ulna** the medial lower arm bone when in anatomic position.
- **Carpals** eight rounded bones of the wrist.
- **Metacarpals** five cylindrical long bones of the palm of the hand.
- **Phalanges** finger bones; two in the thumb and three in each of the other fingers.
- **Digit** a whole finger.

Hip and Leg

Observe the hip and leg bones listed below.

- **Pelvis** the large "hip" bone, consisting of the superior ilium, posterior/interior ischium and anterior/inferior pubis.
- **Femur** the single large upper leg bone.
- **Patella** the kneecap.
- **Tibia** the medial lower leg bone, or shin bone.
- **Fibula** lateral lower leg bone.
- **Tarsals** the seven rounded "ankle" bones.
- **Metatarsals** five cylindrical long bones of the sole of the foot.
- **Phalanges** two in the big toe and three in each of the other toes.

Male vs. Female Pelvis

Observe the male pelvis as compared to the female pelvis (Figure 6.9). Notice that the male pelvis is narrower and funnel-shaped, while the female pelvis is wider, shallower, and basin-shaped. Also note that the male coccyx curves forward, while the female coccyx is more vertical.

Why do you think these differences occur between the sexes?

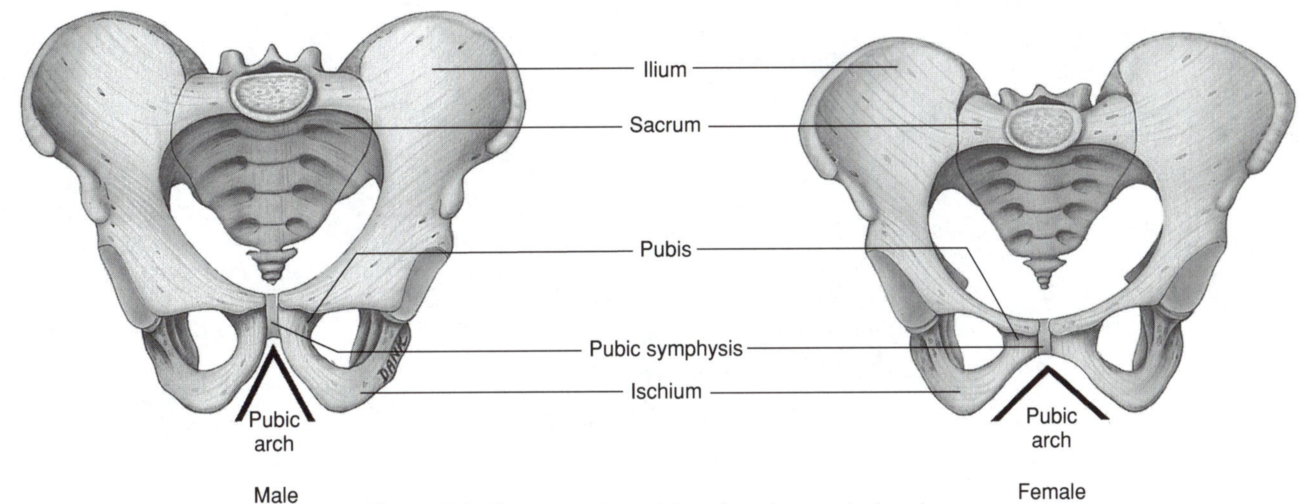

Figure 6.9 Human male and female pelves: anterior view.

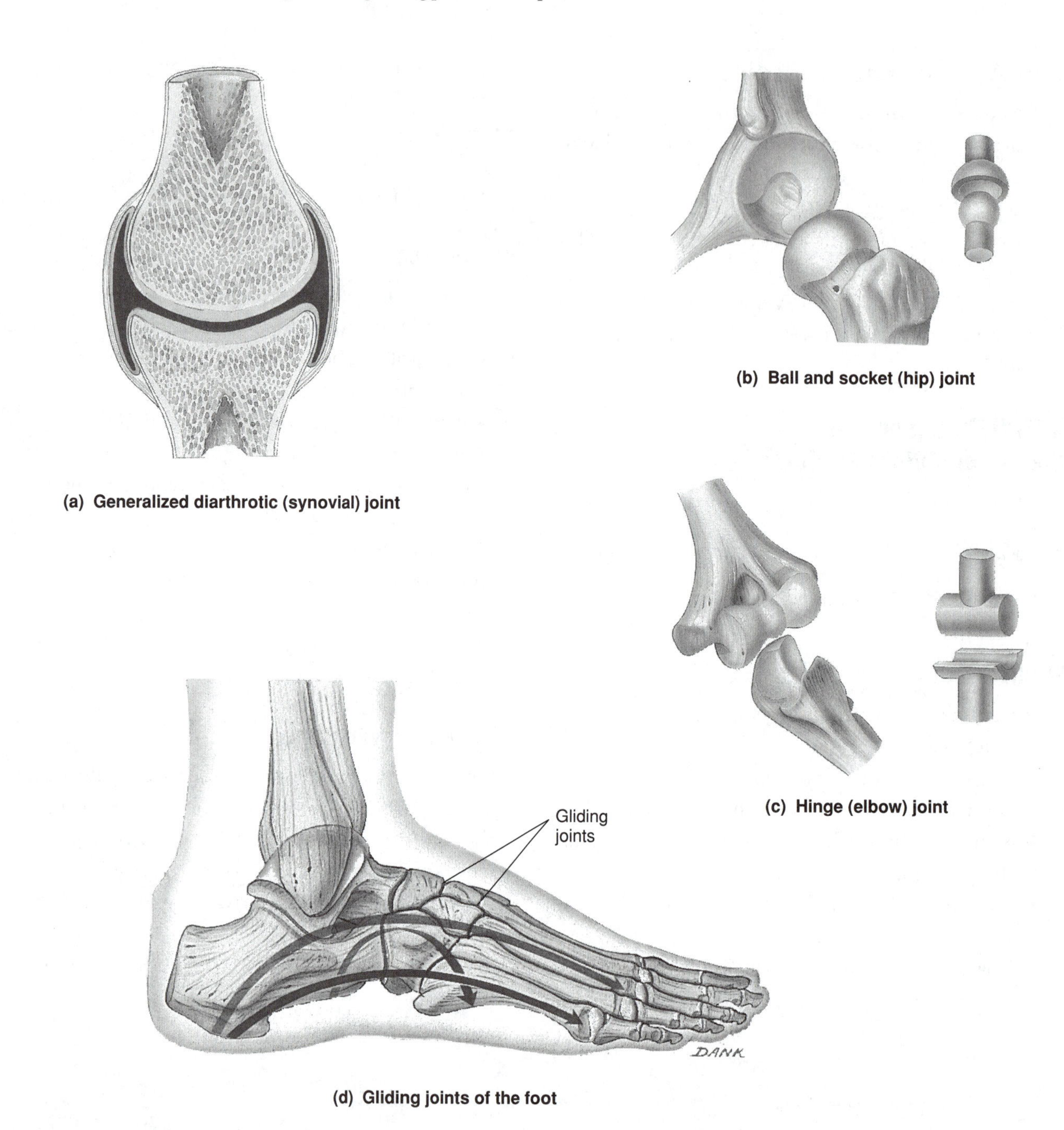

(a) Generalized diarthrotic (synovial) joint

(b) Ball and socket (hip) joint

(c) Hinge (elbow) joint

(d) Gliding joints of the foot

Figure 6.10 Diathrotic (freely moveable) joints.

Joint Types

Now that you are familiar with major bones of the body, use Figure 6.5 and 6.10 to help you locate some major joint types of the body. Observe the joints as categorized below.

Freely Movable

Freely movable joints have synovial cavities and fluid for lubrication.

- **Ball and socket** the "ball" end of one bone fits into a "socket" of another as in the shoulder (humerus) and hip (femur).
- **Ovoid** an egg-shaped surface of bone fits into a reciprocal-shaped surface on another, as in the wrist joint between the carpals and the ends of the radius and ulna.
- **Hinge** freely movable, one plane only (as in the knee and elbow).
- **Pivot** joint spins around an axis as between atlas and axis vertebrae.
- **Gliding** nearly flat bone surfaces allowing side to side and back and forth motion, as between the carpals of the wrist and tarsals of the ankle.

Slightly Movable

Slightly movable joints have no synovial cavity and permit only slight movement. They have a pad of fibrocartilage between two bones.

- **Pubic symphysis** the cartilage connection between the pubic bones.
- **Intervertebral discs** slightly movable joints between most unfused vertebrae other than the atlas and axis.
- **Sacroiliac** between the sacral bone and the ilium of the pelvis.

Immovable

These bones are fitted together by an interlocking arrangement and are held together by fibrous connective tissues. The sutures between the skull bones are the best examples. ■

Study Questions

1. Outline the two major divisions of the human skeleton.

____

____

2. Define the following:

a. diaphysis

____

____

b. epiphysis

____

____

c. medullary cavity

____

____

d. osteocyte

____

____

e. lacuna

____

____

3. Give at least two major differences between the human male and female pelves.

____

____

____

____

4. What are the two major types of bone tissues within a whole bone? What are their functions?

____

____

____

____

5. In which joint-type categories do each of the following belong?

   a. Most vertebrae

   ____

   b. Sutures

   ____

   c. Shoulder-humerus

   ____

   d. Sacroiliac joint

   ____

   e. Knee and elbow

   ____

6. What is a Haversian system?

____

____

____

7. What is unusual about compact bone and cartilage tissue as compared to most body tissues?

____

____

____

8. Write the anatomical name for the following bones.

   a. Second cervical vertebra

   ____

   b. Last two pairs of ribs

   ____

   c. Vestigial tailbone

   ____

   d. Shoulder blade

   ____

   e. Upper jawbone

   ____

9. Name the three ear ossicles (common and Latin names).

____

____

____

____

10. List at least four major functions of the human skeleton.

____

____

____

____

____

____

EXERCISE 7

The Muscular System

OBJECTIVES

After completing this exercise, you should be able to:

1. Understand the basic function of muscle cells.
2. Recognize the three basic types of muscle tissue and understand their function.
3. Understand the basis for body movements produced by the skeletal muscles.
4. Locate and demonstrate the action of selected skeletal muscles.
5. Understand the macrostructure and microstructure of muscle from whole muscle to the sarcomere.
6. Understand the fundamentals of muscular stimulation and the neuromuscular junction.
7. Understand the basics of muscle physiology.

MATERIALS

- ❑ Microscope

Prepared slides:

- ❑ Skeletal muscle
- ❑ Cardiac muscle
- ❑ Smooth muscle

- ❑ Models of whole muscle, muscle fiber, and the sarcomere
- ❑ Human torso models
- ❑ Human arm and leg muscle models
- ❑ Charts diagramming the muscular system
- ❑ Textbook
- ❑ Tape measure
- ❑ Dynamometer
- ❑ Sarcomere model

Introduction

Of all tissues in the human body, muscle is second only to connective tissue in its proportion to total body mass, and it is the only tissue capable of producing movement. A muscle, or combination of muscles, is responsible for all body movements—both voluntary and involuntary. Walking, writing, swallowing, breathing, making facial expressions, and moving food through our digestive system all occur as a result of muscle contraction. The cells of this tissue are specialized to shorten when stimulated (contraction), and will return to their original length as stimulation ends if they are stretched (relaxation). A muscle cannot "un-contract," but must be pulled back to its original length by gravity or by opposing (antagonistic) muscles. Although their ability to change shape is unique among body tissues, muscle tissues share with nervous tissue the ability to respond to and conduct nervous impulses.

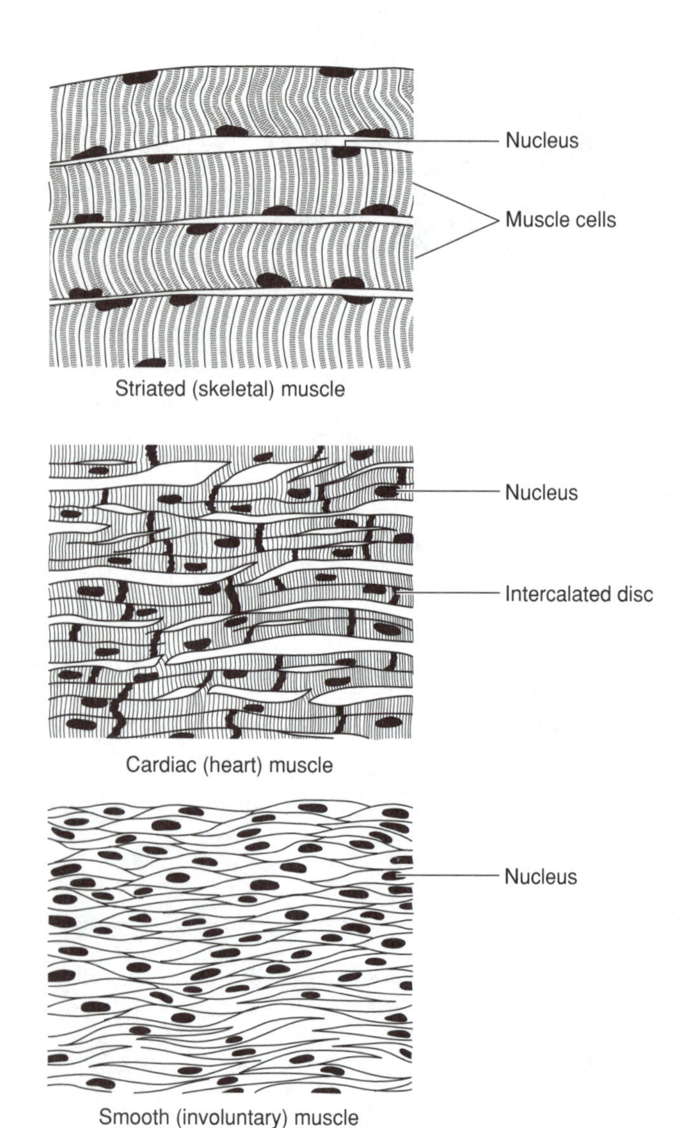

Figure 7.1 Three types of muscle tissue.

There are three very different types of muscle tissue found in the human body: **skeletal, cardiac,** and **smooth** muscle tissues (Figure 7.1). Most of this exercise will be concerned with skeletal muscles.

Skeletal Muscle Cells

Skeletal muscle cells, also called striated or **voluntary** muscle cells, are very unusual when compared with all other cells in the human body. They are very long (sometimes over one foot long) and threadlike, and are multi-nucleated. When viewed under the microscope, they appear to have cross-stripes or **striations.** These striations are really just alternating light and dark areas inside the cell due to the tightly packed pattern of contractile protein fibers, or **myofilaments,** found there. The myofilaments **actin** and **myosin** slide past each other, similar to the way parts of a telescope or radio antenna slide, thus allowing the muscle cells to shorten. This active shortening is generally used to produce movement by taking advantage of the levers provided by the skeletal system.

Skeletal muscle must be stimulated by a neuron (nerve cell) in order to initiate contraction. The **neuromuscular junction** (or myoneural junction) is the area where the neuron branch (axon) ends at a muscle cell (Figure 7.2). The **motor end plate** is a specialized region of the muscle cell membrane. The axon and muscle cell do not actually touch. The "impulse" is not passed on to the muscle by direct connection, but by a chemical **neurotransmitter** released from the synaptic knob that stimulates the muscle cell to contract. Because one axon of a neuron branches to a number of muscle cells, the term **motor unit** is used to describe the motor neuron and all of the muscle cells it stimulates. An axon may branch extensively to many muscle cells where strength, not precision, is needed (i.e., most leg muscles), or it may branch very little if fine control is needed (eyes and fingers).

Cardiac Muscle Cells

Cardiac (heart) muscle cells also appear lightly striated, but unlike skeletal muscle cells, cardiac muscle cells are branched and interconnected. This branching, along with specialized areas of connection between cells called intercalated discs, allow the heart to function as a hollow muscle pump. These characteristics are important where it is critical that the stimulus for contraction spreads throughout the entire heart muscle.

Smooth Muscle Cells

Smooth muscle cells are shorter (not fiberlike) and lack striations completely. They are spindle-shaped,

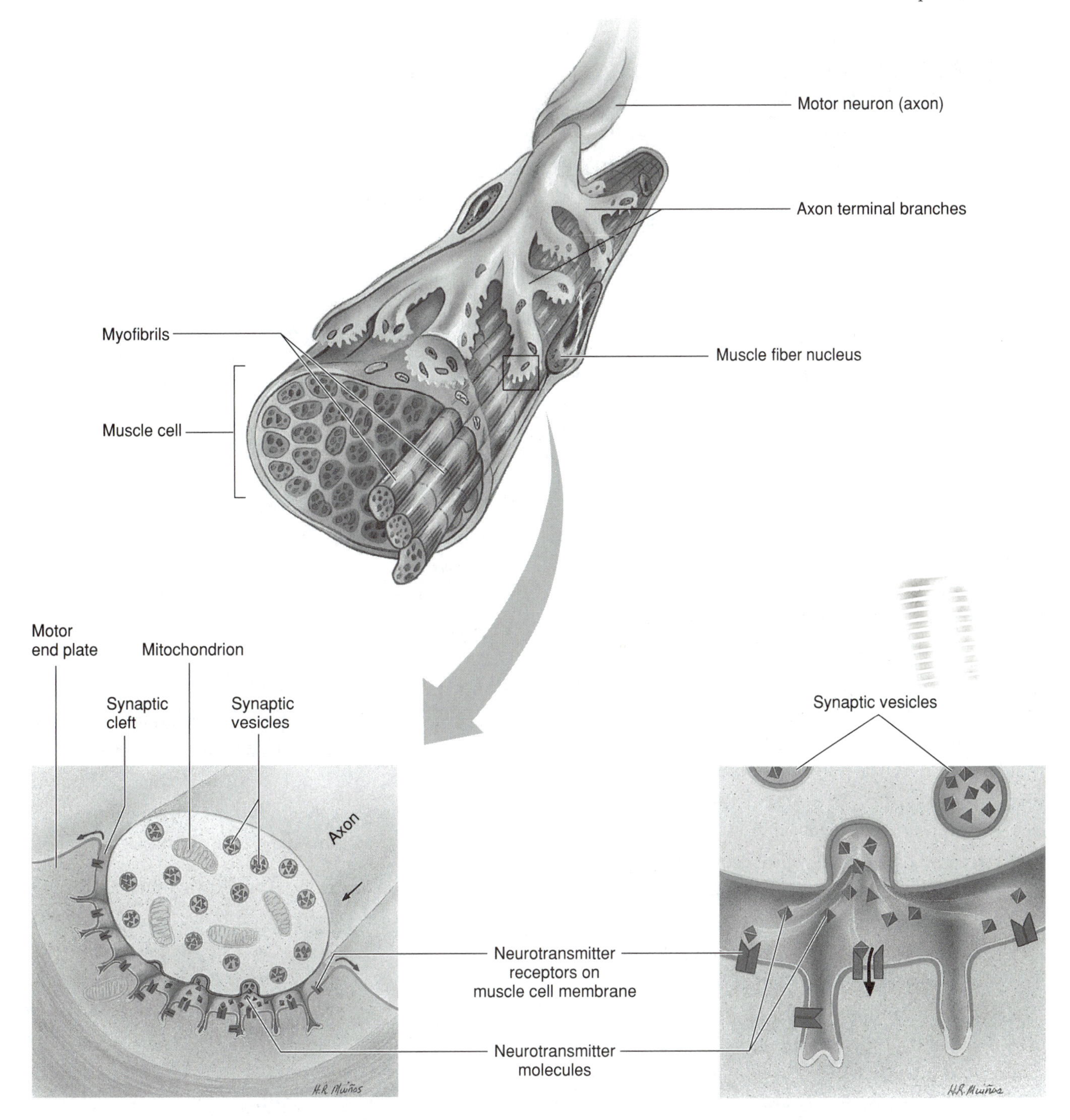

Figure 7.2 Motor unit and neuromuscular junctions.

like an ear of corn—thick in the middle, tapered at the ends. Smooth muscle is the involuntary muscle tissue found in most internal organs requiring movement (i.e., the stomach, intestine, blood vessels, and uterus).

ACTIVITY: Microscopic Examination of Muscle Tissue

See Figure 7.1 for a comparison of the three types of muscle cells—skeletal, cardiac, and smooth. Observe all three examples of muscle tissue and the slide of the myoneural (neuromuscular) junction under the microscope. ■

Skeletal Muscles

Skeletal muscle cells are packed with contractile proteins. These **actin** and **myosin** proteins, or **myofilaments**, slide past each other using the energy from ATP molecules to pull them along. The sliding filament model of muscle contraction states that small proteins on the myosin myofilaments grab the actin proteins and pull them together. (Consult your textbook for more information on this process.) The pattern created by this tight packing of actin and myosin produces alternating light and dark areas called **striations.** The dark areas where the thick myosin myofilaments are found are called **A-bands**, and the lighter areas where only the thinner actin myofilaments are found are called **I-bands** (see Figure 7.3). The **Z-line** represents the thin dark line where sets of actin myofilaments are woven together. The space in between two Z-lines represents a **sarcomere**, or functional unit of contraction.

Muscle cells are wrapped by layers of connective tissue to produce the organ we commonly call a **muscle** (Figure 7.3). Microscopically, each muscle cell is wrapped by a delicate layer of connective tissue called the **endomysium.** The **perimysium** then wraps several muscle cells together into a bundle called a **fascicle.** The outer **epimysium** surrounds the entire muscle and tapers down to a **tendon,** which attaches the muscle to a bone.

Assuming the muscle spans a movable joint, one bone will generally be pulled toward the point of attachment to another bone. These points of attachment are called the **origin** (usually the stationary point of attachment for most actions) and **insertion** (the bone that usually moves toward the origin for most actions). The description of the movement occurring during contraction is the **action** or function of the muscle. Very precise terminology must be used when describing the action of a muscle, and the most common terms are defined here.

Terminology

- **Flexion** decreasing the angle of a joint, as in bending the elbow.
- **Extension** increasing the angle of a joint, as in straightening the elbow joint.
- **Hyperextension** extension beyond the normal anatomical position.
- **Abduction** movement *away from* the midline of the body as in raising the arms laterally.
- **Adduction** movement *toward* the midline of the body, as in lowering the arms from a lateral position.
- **Circumduction** Movement around and away from a particular axis, typically occurring at ball-and-socket joints. For example, you can circumduct your humerus' ball-and-socket joint by drawing a circle on the blackboard with your elbow straight.
- **Rotation** movement around an axis without any lateral displacement, as in shaking your head "no."
- **Supination** turning the palm of the hand upward; or moving the body in anatomical position such that the palms are up (lying face up).
- **Pronation** turning the palm of the hand downward; or moving the body (in anatomical position) such that the palms are down (lying face down).

Muscles are typically grouped together and sometimes named for their type of movement. A **levator** raises and a **depressor** lowers a body part. A **sphincter** closes an opening. A **flexor** decreases the angle of a joint (as in bending the arm), and an **extensor** increases the angle of a joint (usually to a straight position).

Although the study of the muscular system sometimes implies that a single muscle causes a particular movement to be executed, coordinated movement by our muscular system more often requires the teamwork of several muscles. In any movement or **action,** there are **prime movers** (or "agonists"), **antagonists** (oppose or slow down the prime mover), and **fixators** (hold anchoring bones steady). Together, these muscles produce movement with the appropriate speed, force, and grace that we often take for granted.

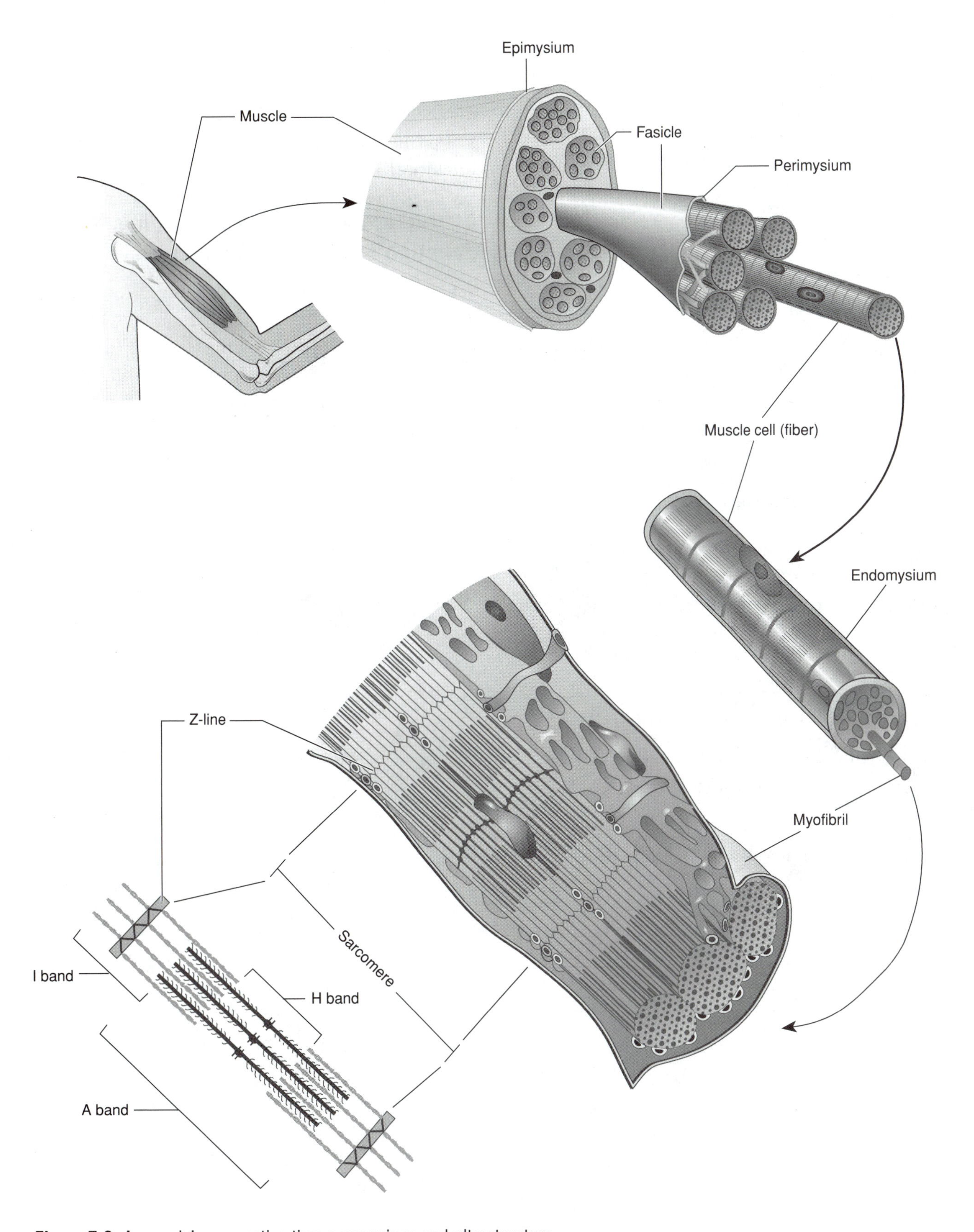

Figure 7.3 A muscle's connective tissue wrappings and ultrastructure.

Without antagonists to slow a prime mover, we would tend to slap a coffee cup rather than reach for it. To understand the importance of the often overlooked fixators, imagine how awkward and unproductive a boxing match would be if the boxers wore roller skates! A solid platform is clearly needed to launch any accurate movement. Holding bones in place and modulating movements are just as important as providing the actual force for the lever.

What are the myofilaments and what is their job in a muscle cell?

____

____

____

Match the listed movement with the term that best describes its action.

________ A. Extension ________ C. Flexion

________ B. Rotation ________ D. Abduction

1. Starting with your chin on your chest, raise your head to look straight ahead.
2. Cross your arms in front of your chest.
3. Sitting with arms at your side and shoulders facing the table, reach for your textbook on the table.
4. Placing your arm out straight, alternate the "thumbs up, thumbs down" position.

Gross Anatomy

It is important to learn about muscles by understanding their *location* (*origin* and *insertion* where attached to skeleton) as well as the kind of movement they produce (*function* or *action*).

ACTIVITY: Learning the Location and Function of Skeletal Muscles

Tables 7.1, 7.2, and 7.3 specify the locations and actions of some of the most important muscles in the body. Be sure that you learn the action or function of these muscles, and identify them on the models. [*Note*: "Arm" means shoulder to elbow; "forearm" is elbow to wrist.] ■

(text continued on page 55)

Table 7.1 Muscles of the Head and Neck (Figure 7.4)

Muscle	Location	Functions
Orbicularis oculi (2)	Encircles each eye	Closes the eyelid, squinting
Orbicularis oris (1)	Encircles the mouth	Closes the lips, puckering
Zygomaticus (2)	Spans from the zygomatic bone to the corner of the mouth	Pulls the corner of the mouth as in smiling
Masseter (2)	Spans from the zygomatic bone to the mandible	Closes the jaw as in chewing
Temporalis (2)	Spans from the temporal bone to the mandible	Closes the jaw as in chewing
Sternocleidomastoid (2)	Diagonals down each side of the neck from behind the ear (mastoid of temporal bone) to the sternum and clavicle	Together, both flex the head; alternately, they rotate the head; each by itself pulls the head laterally
Trapezius (1)	Large diamond-shaped muscle of the upper back which attaches to the occipital, both scapulas, lateral clavicles, and vertebrae	Extends the head; raises and lowers the shoulder as in "shrugging"

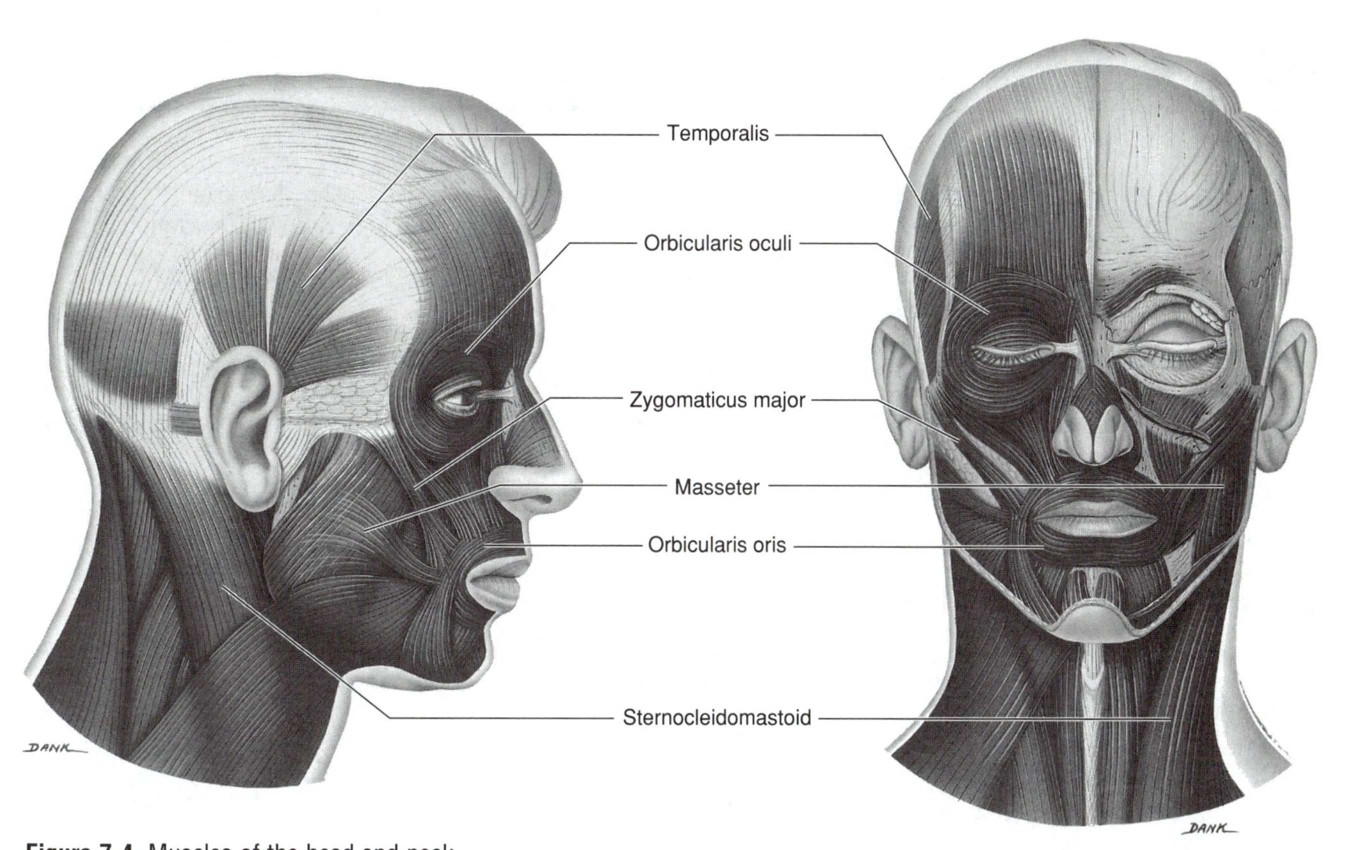

Figure 7.4 Muscles of the head and neck.

Table 7.2 Muscles of the Torso (Figure 7.5a, b)

Muscle	Location	Functions
Latissimus dorsi (2)	Large middle-back muscle that fans out laterally from vertebrae toward the armpit, then attaches to the humerus	Adducts the arm (humerus); extends the arm when in anatomical position (in front of the torso)
Pectoralis major (2)	Large muscle under the breast area; spans mainly from the sternum to the humerus	Adducts the arm (with the latissimus dorsi); flexes the arm (humerus) as in moving the arm in front of the chest
Rectus abdominis (1)	Central abdominal muscle attached to the lower ribs and sternum and the pubis	Flexes the torso (as in doing a sit-up); reinforces the abdominal wall
Gluteus maximus (2)	Spans mainly from the posterior and lateral pelvis to the upper femur	Extends and hyperextends the thigh (femur) as in kicking backwards; abducts thigh
Deltoid (2)	Forms the rounded area of the upper arm and shoulder; connects the scapula and clavicle to the humerus	Mainly abducts the arm (humerus)

Table 7.3 Muscles of the Appendages (Figure 7.5a, b)

Muscle	Location	Functions
Biceps brachii (2)	Spans the entire anterior humerus; connects the scapula and upper humerus to the radius	Flexes arm at the elbow joint (this is also referred to as flexing the forearm)
Triceps brachii (2)	Spans the entire posterior humerus; connects the scapula and upper posterior humerus to the ulna	Extends the arm (forearm) at the elbow joint
Quadriceps femoris group (2) –rector femoris, vastus lateralis, vastus intermedialis (not shown on figure), vastus medialis	Large muscle group of the anterior thigh (femur); connects pubis and upper femur to the tibia	Extends the lower leg at the knee joint; flexes the thigh (femur); both actions together allow “kicking” forward
Hamstring group (2)–biceps femoris, semitendinosus, semimembranosus	Large muscle group of the posterior thigh; connects the posterior pelvis (ischium) to the tibia	Flexes the lower leg at the knee joint; also extends the thigh (femur) when the leg is already raised
Gastrocnemius (2)	Large posterior lower leg or “calf” muscle; connects the lower femur to the heel	Extends the foot as in standing on the toes; assists with flexing the knee joint

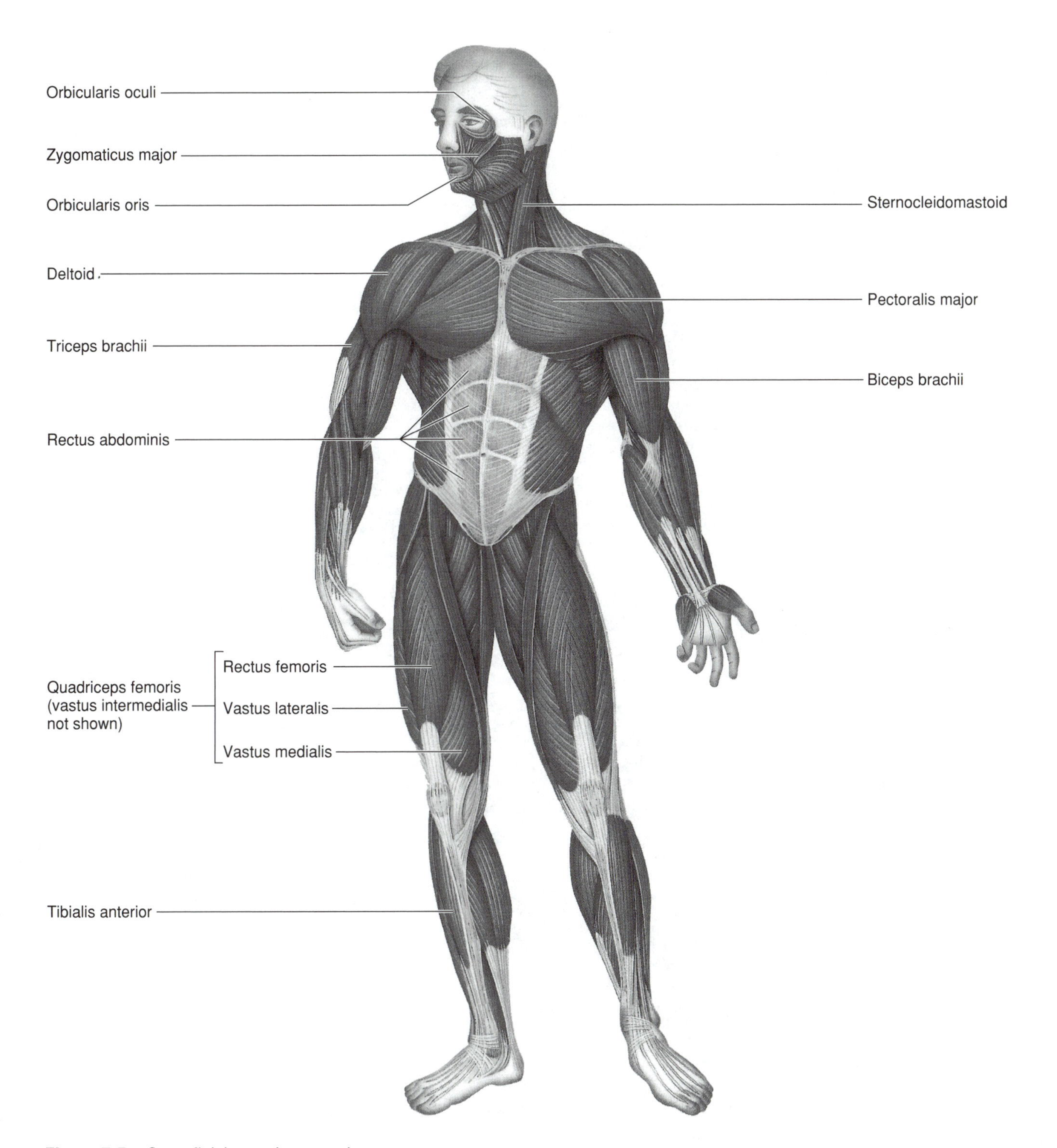

Figure 7.5a Superficial muscles: anterior.

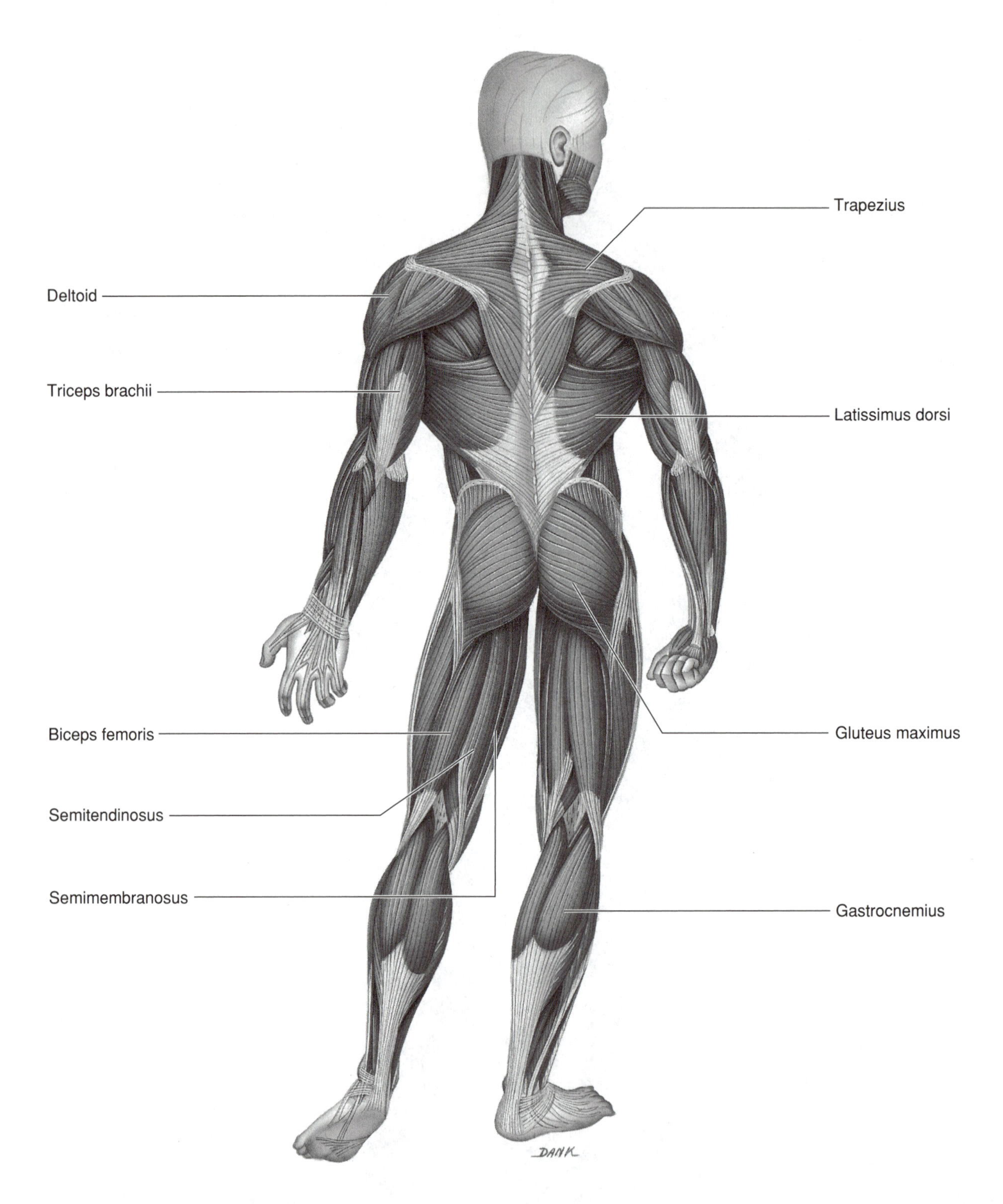

Figure 7.5b Superficial muscles: posterior.

ACTIVITY: Identifying Skeletal Muscles of the Body

On the models available in the laboratory, find the muscles listed in the preceding tables.

1. Intramuscular injections are typically given in the shoulder and buttocks areas. Which muscles are injected?

   ____

   ____

2. Which muscle is the prime mover for each of the following movements?

   Winking ______________________________________

   Shaking your head "no"________________________

   Hugging ______________________________________

   Jumping ______________________________________ ■

Muscle Physiology

The sliding filament theory suggests that enormous amounts of ATP must be consumed every time a muscle contracts. **Aerobic respiration** is a highly efficient form of ATP production that occurs in the mitochondria of a cell. In order to perform aerobic respiration, most muscle cells have a much higher than average number of mitochondria, as well as stores of glycogen (a carbohydrate) and an oxygen-storing molecule called **myoglobin** (a relative of the hemoglobin molecule). Although some types of muscle cells are more energy efficient than others, most cells, with the exception of cardiac muscle cells, will deplete these reserves when exercised vigorously.

Once oxygen reserves are depleted, cells can still produce ATP, but only using an incomplete and inefficient chemical process. This **glycolysis** or **anaerobic respiration** ends with the production of lactic acid in human muscle cells. Lactic acid is irritating to muscle cells and causes the soreness we associate with excessive muscle use. Additionally, lactic acid makes the vessels feeding skeletal muscles dilate or expand, so that increased blood flow can replace the depleted oxygen and nutrients.

When a muscle cell depletes its ATP much faster than it can be replaced, **fatigue** sets in. Even though the muscle is being stimulated, it cannot respond without the ATP necessary to move those actin and myosin filaments.

Terminology

- **Tonus** the mild state of contraction due to alternate contraction of different motor units (especially important in maintaining posture).
- **Fatigue** inability of a muscle to contract due to the depletion of ATP faster than it can be replaced.
- **Flaccidity** limpness due to lack of tonus.
- **Rigor mortis** muscular stiffness after death due to total loss of ATP and subsequent calcium leakage throughout the muscle. Calcium causes the myofilaments to bind together, and they remain stuck without ATP to facilitate their detachment.
- **Isometric contraction** ("same measure") the stimulus results in tension applied without movement of the muscle or joint.
- **Isotonic contraction** continual tension results in movement of the joint and shortening of the muscle.
- **Tetanus** complete, sustained contraction due to rapid stimulation.

ACTIVITY: Studying Muscle Physiology

Experiment I

1. Measure the circumference of your left biceps brachii muscle in both the contracted and relaxed state.

   Contracted: ___________ Relaxed: ___________

2. While holding a large textbook, dumbbells, or some other reasonably heavy object, contract and relax the biceps brachii repeatedly until it fatigues. Remeasure the muscle in both states.

   Contracted: ___________ Relaxed: ___________

3. If the numbers are the same before and after, repeat until a difference is obtained.

4. Explain your results below. (*Hint:* your muscle cells did not begin growing that quickly after exercise.)

   ____

   ____

   ____

Experiment II – Dynamometer (Gripper) Experiment

Part A

1. Squeeze the gripper three times with as much force as is comfortable. Record the average of your measurements.

   ____

2. Reset the gripper; rest for a few seconds; repeat and record this second average.

   ____

3. Which average was higher?

   ____

   Why?

   ____

   ____

   ____

Part B

1. Read this entire section before starting, since the procedure must be done quickly to obtain the correct results.

2. The same person as in the previous experiment, using the same hand, should rapidly squeeze the gripper 20 times and read aloud the total force measured for the lab partner to record.

   ____

3. *Immediately* reset and squeeze the gripper another 20 times *without* any rest period in between. Record.

   ____

4. *Immediately* repeat a third time *without* rest. Record.

   ____

5. Which average was the lowest?

   ____

   Why?

   ____

   ____

   ___________________________________ ■

Study Questions

1. Complete Chart 7.1:

Chart 7.1

Muscle Cell Type	General Appearance	Location	Voluntary / Involuntary	Fast or Slow
Skeletal				
Smooth				
Cardiac				

2. Name the smallest contractile unit of striated muscle.

____

3. Make a sketch of a sarcomere and label at least five items.

4. Briefly explain what happens when a muscle contracts.

____

____

____

____

____

____

5. Define tonus.

6. Make a sketch of the neuromuscular junction and label at least three items.

7. Which type of muscle does not consist of fiberlike cells?

8. Fill in the term that best fits the description given:

   a. Moves a joint toward the midline of the body.

   b. Raises a body part.

   c. Protein that composes the thick filament.

   d. Any muscle that bends decreases the angle of a joint.

   e. Tension applied without shortening.

   f. Muscular stiffness due to death.

   g. Mild state of contraction for most muscles.

9. Name the muscles described below.

   a. Shoulder muscle that abducts the arm.

   b. Large diamond-shaped muscle that extends the head.

   c. The two muscles involved in chewing.

   d. Posterior humerus muscle that extends the forearm.

   e. Large middle-back muscle that adducts and extends the arm.

   f. Calf muscle that extends the foot.

10. Name the two muscle groups that move the lower leg and state their main actions.

EXERCISE 8

Nervous System I: *Organization, Neurons, Spinal Cord, Spinal Nerves, and Spinal Reflexes*

OBJECTIVES

After completing this exercise, you should be able to:

1. Describe the organization of the nervous system.
2. Name the types of neurons and neuroglial cells.
3. Describe and identify the structures of the neuron using models and slides.
4. Discuss the functions of neuron structures.
5. Describe the meninges of the brain and spinal cord.
6. Describe the structure and function of the spinal cord.
7. Identify the components of the peripheral nervous system.
8. Describe the relationship between neurons, axons, and nerves.
9. Name the structures found in a nerve.
10. Explain what a plexus is, name and identify the somatic plexuses, and identify a major nerve emerging from each plexus.
11. Define a reflex and give examples of reflexes involving cranial nerves, spinal nerves, and the ANS.

MATERIALS

- ❑ Model of the neuron with model key
- ❑ Models of the spinal cord
- ❑ Spinal cord model showing spinal nerves and plexuses
- ❑ Slide: cross-section of spinal nerve

Prepared slides:

- ❑ Medullated neuron (demo)
- ❑ Neurons (ox smear)
- ❑ Cross-section of spinal cord (demo)

Introduction

The nervous system is our most rapid means of maintaining homeostasis. Using a combination of electrical and chemical signals, the nervous system can:

- detect changes in both the external and internal environment.
- interpret these environmental changes.
- initiate responses to these environmental changes in the form of skeletal, smooth, or striated muscle contractions and glandular secretions.

The organization of the nervous system is represented in Figure 8.1.

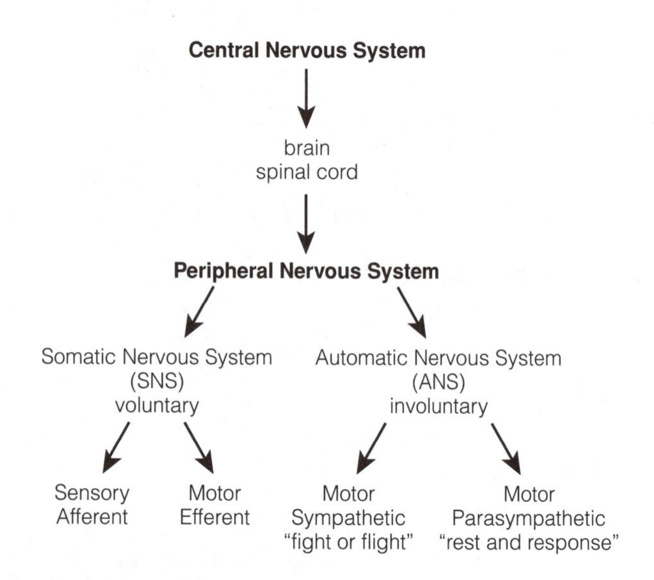

Figure 8.1 Organization of the nervous system.

Histology of the Nervous System

The nervous system is composed of two very different types of tissue. **Neurons** are the conductive cells that transmit messages, and **neuroglia** are the supporting cells of the nervous system.

Neurons

All neurons are cells specialized to conduct impulses. Neurons may be classified based on structure or function. Regardless of function, all neurons generally have a **cell body** which contains the nucleus, an **axon** that can often carry impulses over great distances, and **dendrites** which are capable of receiving stimuli.

Structural Classification

Unipolar Neurons

All unipolar neurons have a single process emerging from a round cell body (Figure 8.2). The process splits. One branch will behave as the **afferent** and the other will behave as the **efferent** branch. All unipolar neurons are sensory (transmit sensation). See "Functional Classification" for more information.

Bipolar Neurons

Bipolar neurons have two processes emerging from a round cell body. They are not common in the body, but are found as a single layer of cells in the retina of the eye and in the ear. All bipolar neurons are sensory.

Multipolar Neurons

The cell body of a multipolar neuron (Figure 8.3) is usually stellate (star-shaped). Many short processes called dendrites emerge from the cell body. They increase the surface area of the neuron available for receiving signals. A single longer process called the axon emerges from the cell body and goes to the effector. Multipolar neurons are motor or association in function.

Functional Classification

Sensory (Afferent) Neurons

Sensory neurons carry sensory information from receptors to the brain and spinal cord. Most are **unipolar**, with a small number of **bipolar** neurons.

Association (Internuncial) Neurons

Located in the CNS (brain or spinal cord), association neurons are the **connecting** link between **sensory** and **motor** neurons. All are **multipolar** neurons.

Motor (Efferent) Neurons

Motor neurons carry messages from the CNS (brain and spinal cord) to the **effectors** (muscles and glands). They cause something to happen (**motion or secretion**). All are **multipolar** neurons.

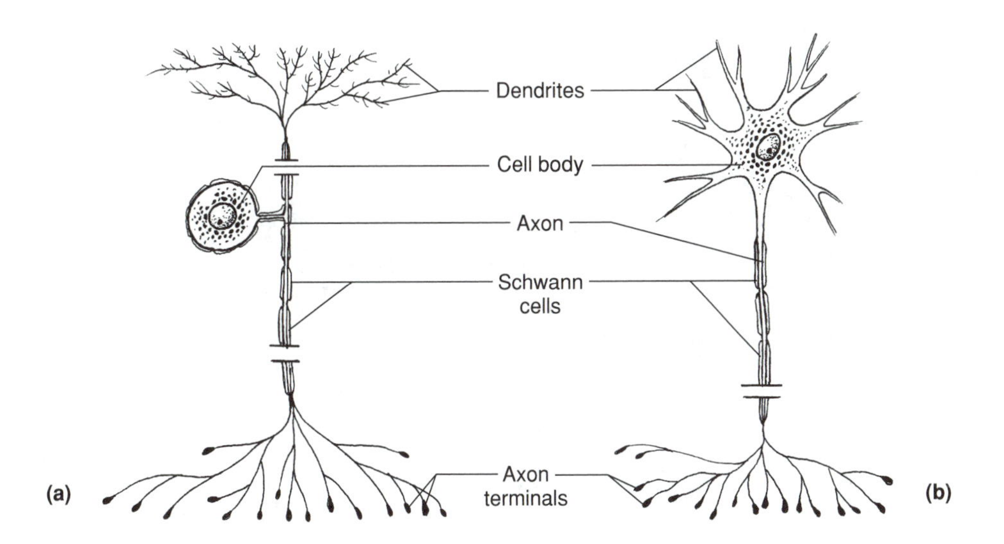

Figure 8.2 Structural classification of neurons: a) Unipolar (sensory) and b) multipolar (motor) neurons.

Neuroglial Cells

Neuroglial ("nerve glue") cells are generally believed to be connective tissue cells. Their function is to support, protect, and nourish neurons. Some cell types and their functions are:

- **Schwann cells** form the protective myelin sheath in the PNS.
- **Oligodendrocytes** form the protective myelin sheath in the CNS.
- **Microglia** phagocytic cells; destroy bacteria and damaged tissues.
- **Astrocytes** contribute to the blood-brain barrier and scar tissue.
- **Ependymal cells** ciliated cells that line the ventricles and move cerebrospinal fluid.

Structure of a Neuron

The neuron has numerous structures specialized for conducting impulses. These structures are outlined below.

Perikaryon (Cell Body)

Unipolar and bipolar neurons have a round cell body. Multipolar neurons have a large **nucleus** and the **stellate** (star) shape. The cell appears to be similar to other body cells, *but* it lacks centrioles.

How might this affect the neuron?

____

____

____

____

Nissl Bodies

Nissl bodies are composed of rough endoplasmic reticulum in which neurotransmitters are synthesized.

Dendrites

Dendrites are short processes that emerge from the perikaryon in a multipolar neuron, or from the peripheral process of a unipolar neuron. They are formed from the plasma membrane and increase the

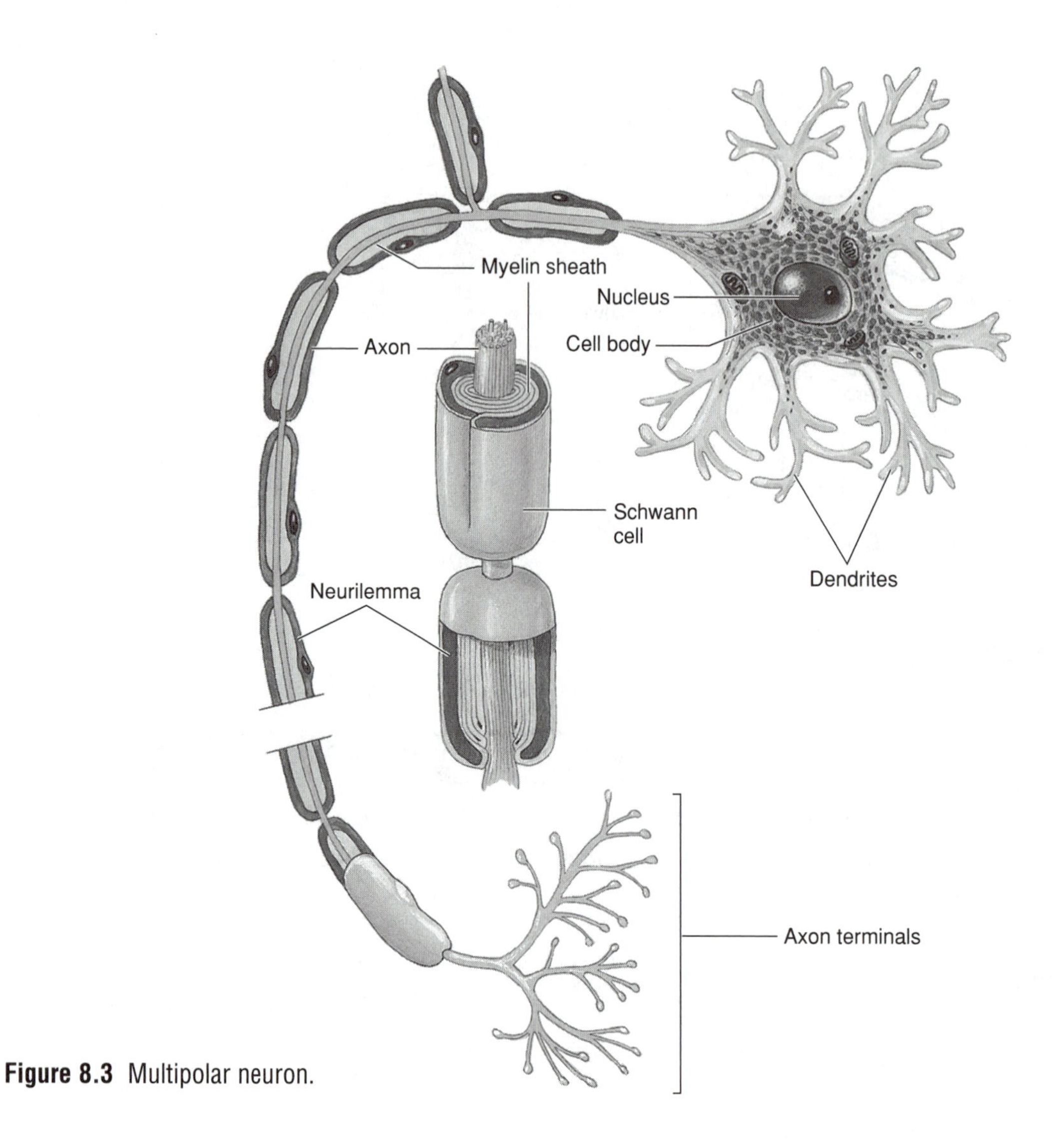

Figure 8.3 Multipolar neuron.

surface area for reception of transmissions from many other neurons. In multipolar neurons, they transmit impulses to the perikaryon and then on to the axon.

Axon

An axon is a single, longer process. The axon emerges from the perikaryon and transmits impulses from the perikaryon to the **effectors.** Axons have branches called **collaterals.** Collaterals may undergo further branching, forming the **axon terminals.** Each axon terminal ends in a **synaptic knob** (bouton, button, or end bulb).

Synaptic Knob (End Bulb)

The synaptic knob converts the **electrical impulse** to a **chemical impulse** by releasing a **neurotransmitter.** When transmission is to a skeletal muscle, the neurotransmitter is **acetylcholine.** Since nerve transmission is extremely complex and specific, many neurotransmitters are needed to provide specificity. There are currently over 90 known transmitters.

Myelin Sheath

The myelin sheath is a fatty wrapping of the axon that provides insulation for the axon. In the PNS many Schwann cells wrap themselves entirely around a short length of axon. Between each Schwann cell wrapping there is a tiny area of unmyelinated axon called the node of Ranvier. In the CNS an oligodendrocyte that has many broad, flat extensions wraps many regions of a single axon.

Node of Ranvier

The node of a Ranvier is a patch of unmyelinated axon. Electrical impulses can occur only in these patches. Thus, the impulse must skip over the myelinated axon going from node to node. Transmission of this type is called **saltatory** and is much more rapid than transmission in unmyelinated neurons.

Neurilemma

When a Schwann cell of the PNS wraps the axon, the initial wrapping is composed of lipid membrane only. It is only the very last layer of wrapping that includes the cytoplasm and nucleus. This last layer of wrapping is called the **neurilemma** and is essential for **axonal repair.** An **oligodendrocyte** (myelin-producing cell of the CNS) may wrap myelin around several places on one axon, but there is *no neurilemma* on myelinated axons of the CNS. Injured PNS axons may be repaired, but injured CNS axons are *never* repaired. The nucleus-containing neurilemma is needed.

ACTIVITY: Identifying the Structure of a Neuron

The most commonly available models and slides are of **multipolar neurons.**

1. Look at the model of a neuron and identify the structures named above. All may be seen on the model.

2. Look at the slide of neurons labeled "ox smear."

   a) Which structures are easily identified?

   ____

   ____

   ____

   b) Which structures cannot be seen on the slide?

   ____

   ____

   ____

   c) Why can't you see all of the named structures?

   ____

   ____

   ____

3. Look at the demonstration slide of a medullated (myelinated) neuron, and explain what you see.

   ____

   ____

   ____

   ____

   ______________________________ ■

Central Nervous System (CNS)

The CNS includes the **brain** and **spinal cord,** which serve to control and integrate all nervous functions. Most of the neuron cell bodies of the nervous system are located in the CNS. Cell bodies and unmyelinated axons impart a **gray color** to brain and spinal cord tissue whereas myelin, a lipid, imparts a **white color** (thus, the origin of the terms **gray** and **white matter**). Since neurons lack centrioles, they cannot reproduce themselves; thus, the brain must be protected from injury.

Protection of the CNS

Protection of the CNS is derived from the *skull and vertebrae* surrounding the brain and spinal cord, *adipose tissue* within the vertebral canal, the *meninges,* three connective tissue wrappings of the CNS, and the cerebrospinal fluid (Figure 8.4).

The **meninges** (singular is meninx) include:

- **Dura mater** "tough mother;" a leathery outermost wrapping composed of connective tissue.
- **Arachnoid** a thin cobwebby middle layer.
- **Pia mater** "soft mother;" a thin but tough connective tissue wrapping which adheres to the surface of the brain and spinal cord.

Cerebrospinal fluid (CSF) is a layer of fluid formed from blood plasma but is considerably lower in protein content. It circulates in the **subarachnoid space** between the **pia mater** and the **arachnoid layer**. Fluid is highly incompressible and, therefore, is an excellent shock absorber.

Where else is fluid used in a similar manner? (*Hint:* think pregnancy)

____

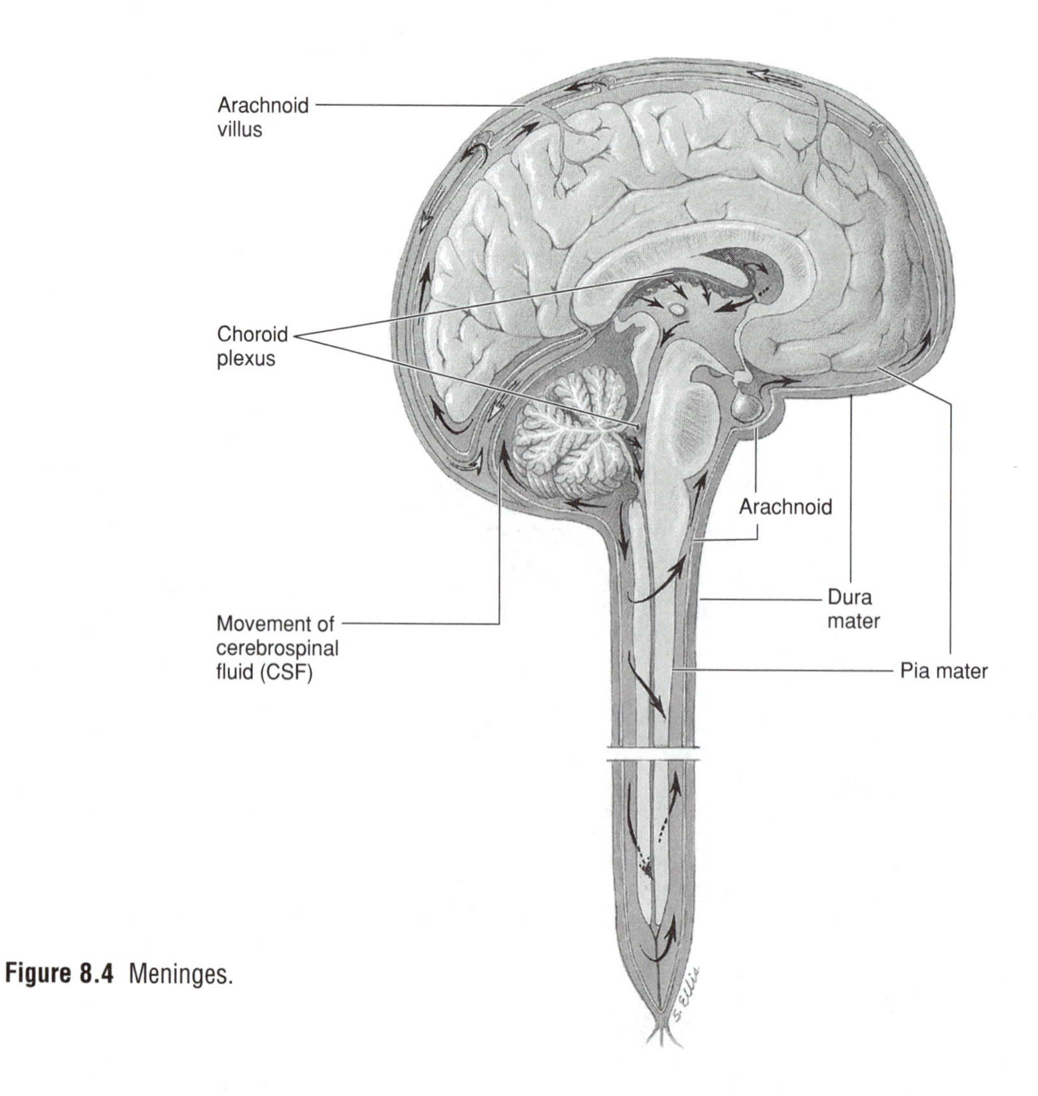

Figure 8.4 Meninges.

CSF can also be found circulating within the brain in the **ventricles** and within the spinal cord in the **spinal canal,** where it aids in providing nutrients for cells and removing waste.

Peripheral Nervous System (PNS)

The peripheral nervous system is composed of *receptors* and *nerves*. The nerves include the 12 pairs of **cranial nerves** emerging from the brain and the 31 pairs of **spinal nerves** emerging from the spinal cord, as well as all of the **nerves of the autonomic nervous system (ANS).**

Organization of a Nerve

A **nerve** is a bundle of axons, often grouped in fascicles just as muscle cells were in skeletal muscle. Most of the cell bodies of neurons are found in the central nervous system or in ganglia near the CNS. The **endoneurium** is the connective tissue wrapping each axon. **Fascicles** are bundles of axons wrapped by a layer of connective tissue called the **perineurium**, connective tissue around the fascicles. The **nerve** is typically a bundle of many fascicles wrapped by a layer of connective tissue called the **epineurium**.

Spinal Cord

Gross Anatomy

The spinal cord (Figure 8.5) is located in the vertebral column and is approximately the diameter of the thumb, although it is not uniform in thickness (it has *cervical* and *lumbar enlargements*). Averaging about 18 inches long the spinal cord terminates at L2. 31 pairs of spinal nerves branch from the spinal cord. The **cauda equina** (horse's tail) are the nerves that span the lumbar region, from the termination of the cord at the **conus medullaris** to their exit points from the lower vertebral column.

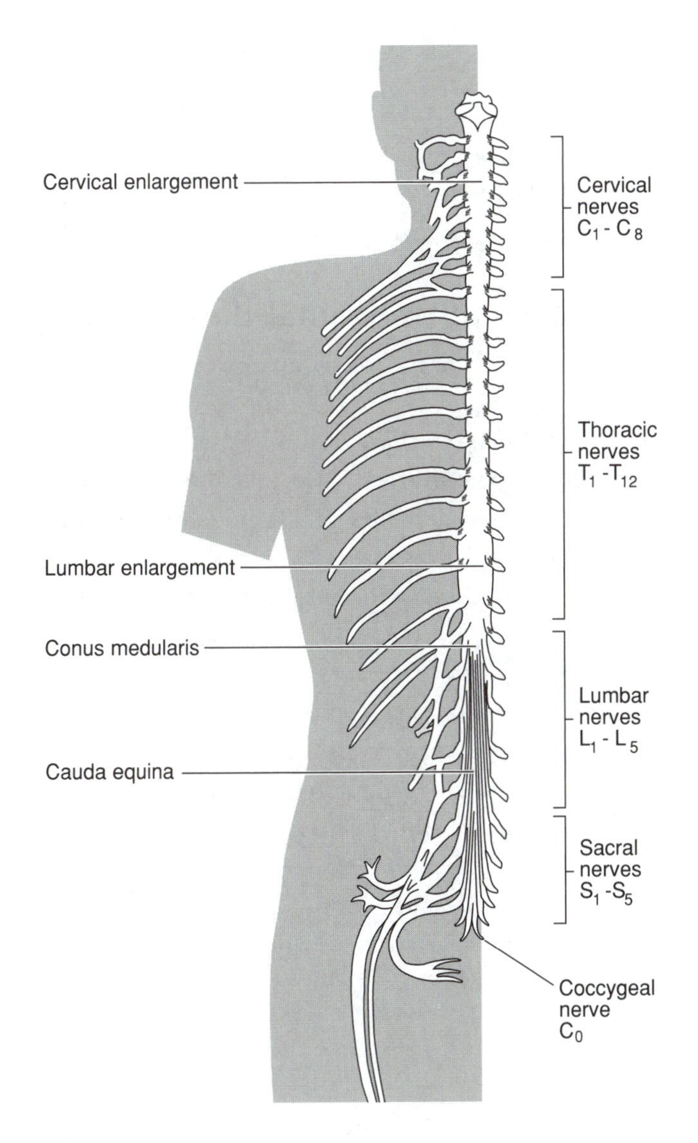

Figure 8.5 Human spinal cord.

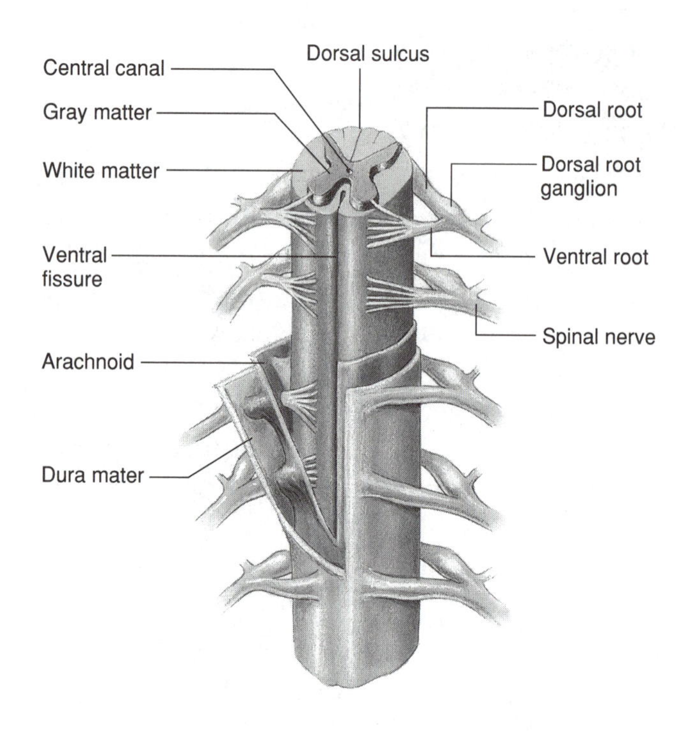

Figure 8.6 Human spinal cord with nerve attachment: cross section.

Histology of the Spinal Cord

The spinal cord has the following microscopic features (Figure 8.6):

- **Central canal** where CSF circulates.
- **Central gray matter** also called gray "H"; location of all cell bodies and unmyelinated axons found within the cord. All synapsis between sensory, association, and motor neurons takes place here.
- **White matter** ascending (sensory) and descending (motor) tracts composed of myelinated axons.
- **Ventral (anterior) fissure** the deep groove located on the ventral side of the spinal cord.
- **Dorsal (posterior) sulcus** a shallow groove on the dorsal side of the spinal cord.
- **Ventral (anterior) root** all axons from motor neurons in the ventral grey horn emerge from the spinal cord in the ventral root.
- **Dorsal (posterior) root** all axons from sensory neurons enter the spinal cord using this pathway.
- **Dorsal root ganglion** all sensory neuron cell bodies are located here.
- **Spinal nerves** dorsal and ventral roots unite to form the spinal nerves. Since they have axons of both sensory and motor nerves, spinal nerves are considered **mixed nerves**.

Functions of the Spinal Cord

The spinal cord conducts impulses to and from the brain and also from one part of the spinal cord to another. It is the receiving point in the CNS for sensory information entering via the dorsal root. It transmits motor information via the ventral root. It is also the center for all spinal reflexes.

What is the effect of severing the spinal cord?

____

____

What is the effect of severing the dorsal root?

____

____

What is the effect of severing the ventral root?

____

____

What is the effect of severing both roots?

____

____

ACTIVITY: Identifying Structures of the Spinal Cord

Use the model of the spinal nerve to identify the structures in Figure 8.6. View slides of the spinal cord that your instructor may have put out as demonstration material. ■

Spinal Nerves

Spinal nerves are all mixed nerves, but the roots are not mixed. Call the structure of the nerve roots as they emerge from the spinal cord. The dorsal root is sensory only and has the dorsal root ganglion which is the location of all of the cell bodies of the sensory neurons.

The ventral root has motor function only. It is composed of axons only.

Where are the cell bodies of these axons?

____

____

____

____

The two roots unite, forming the mixed spinal nerve which may unite with other nerves to form a plexus (Figure 8.7). A **plexus** is a tangle of nerves. Nerves entering may leave the plexus with fibers from other nerves. Spinal nerves do not have names. They are assigned an initial for the region of the spinal cord and a number.

There are eight pairs of **cervical nerves,** C1 through C8. Most go to the **cervical plexus**, but C8 goes to the brachial plexus. The major nerve emerging from the cervical plexus is the **phrenic nerve,** which innervates the diaphragm.

The **brachial plexus** is formed from C8, T1, and T2. The major nerves to emerge from the brachial plexus are the **radial** and **ulnar nerves**, which innervate the extensor and flexor muscles of the forearm.

There are twelve pairs of **thoracic nerves**, T3 through T12. They do not go to a plexus. They innervate the intercostal muscles used for breathing.

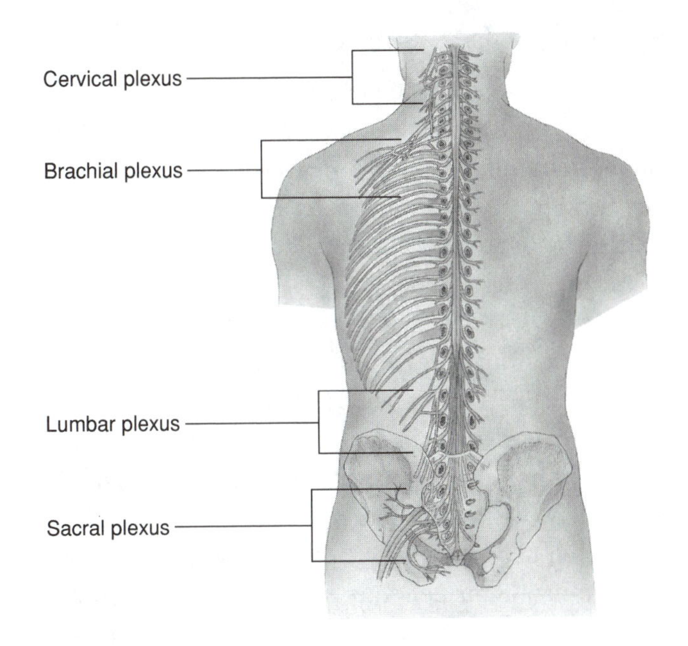

Figure 8.7 Plexuses of the spinal nerves.

There are five pairs of **lumbar nerves**, L1 through L5. Most go to the **lumbar plexus**. The major nerve to emerge from the lumbar plexus is the **femoral nerve,** which innervates the anterior thigh muscles and skin.

There are five pairs of **sacral nerves**, S1 through S5. They go to the **sacral plexus**. The major nerve to emerge from the sacral plexus is the **sciatic nerve,** which innervates the posterior thigh muscles and skin. It is the largest nerve in the body.

The **coccygeal nerve**, Co1, does not go to a plexus. It innervates genitoanal structures.

ACTIVITY: Observing the Spinal Nerves

Refer to Figures 8.4, 8.5, 8.6, and 8.7 and use the models to observe the spinal nerves emerging from the spinal cord and forming the plexuses. We will test the function of selected spinal nerves after we discuss the reflex arc. ■

The Reflex Arc

Reflex arcs are the simplest type of nerve pathway (Figure 8.8). Reflexes are rapid responses to environmental stimuli that are **automatic, unconscious, protective,** and **homeostatic**. Both the spinal cord *and* the brain are reflex centers. Reflex arcs include the following components:

- **Receptor** located at the very beginning of the sensory neuron. It responds to a stimulus and transmits to the sensory neuron.
- **Sensory neuron** stimulated by the receptor, causing an impulse to travel to the brain or the spinal cord. A neurotransmitter is released at a synapse located between the synaptic knobs of the sensory neuron and the dendrites of the association neuron.
- **Association neuron** stimulated by the neurotransmitter, causing an impulse to travel down the axon. A neurotransmitter is released at a synapse located between the synaptic knobs of the association neuron and the dendrites of the motor neuron.
- **Motor neuron** stimulated by the neurotransmitter, causing an impulse to travel down the axon to the synaptic knobs. A neurotransmitter is released at a synapse preceding the effector.
- **Effector** a gland or muscle (smooth, striated, or cardiac) causing secretion or contraction. If the effector is striated muscle, then the transmitter is ACh, which is released at the myoneural junction and stimulates the sarcolemma.

Examples of Spinal Reflexes

Two-Neuron Reflex Path—Patellar Reflex

This reflex involves a sensory and motor neuron only (Figure 8.9). The **receptor** is a stretch receptor in the patellar tendon. When stimulated, it transmits an impulse by way of the **sensory neuron,** which enters the spinal cord and stimulates the **motor neuron.** The motor neuron sends an impulse to the **effector,** or the quadriceps muscle. The end result is extension of the lower leg (the little "kick"). The purpose of this reflex is to compensate for changes in pressure on the leg and enhance coordination during standing and walking.

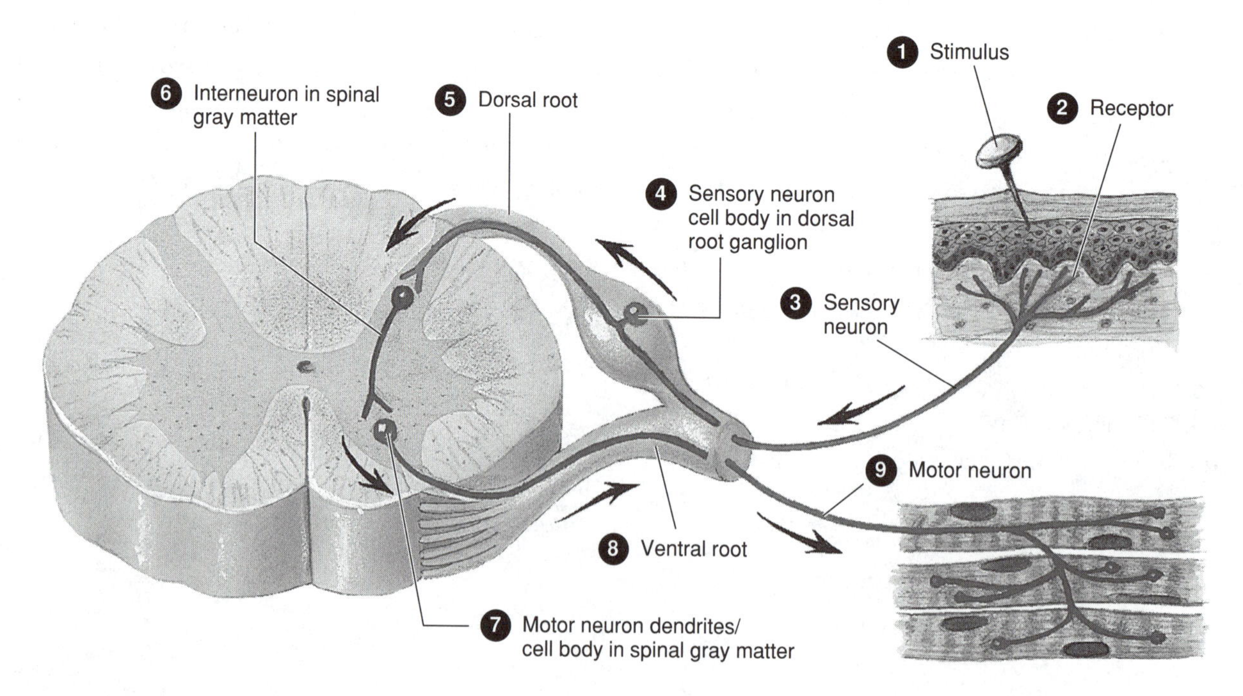

Figure 8.8 General reflex arc. The numbered steps indicate the pathway of a general reflex arc.

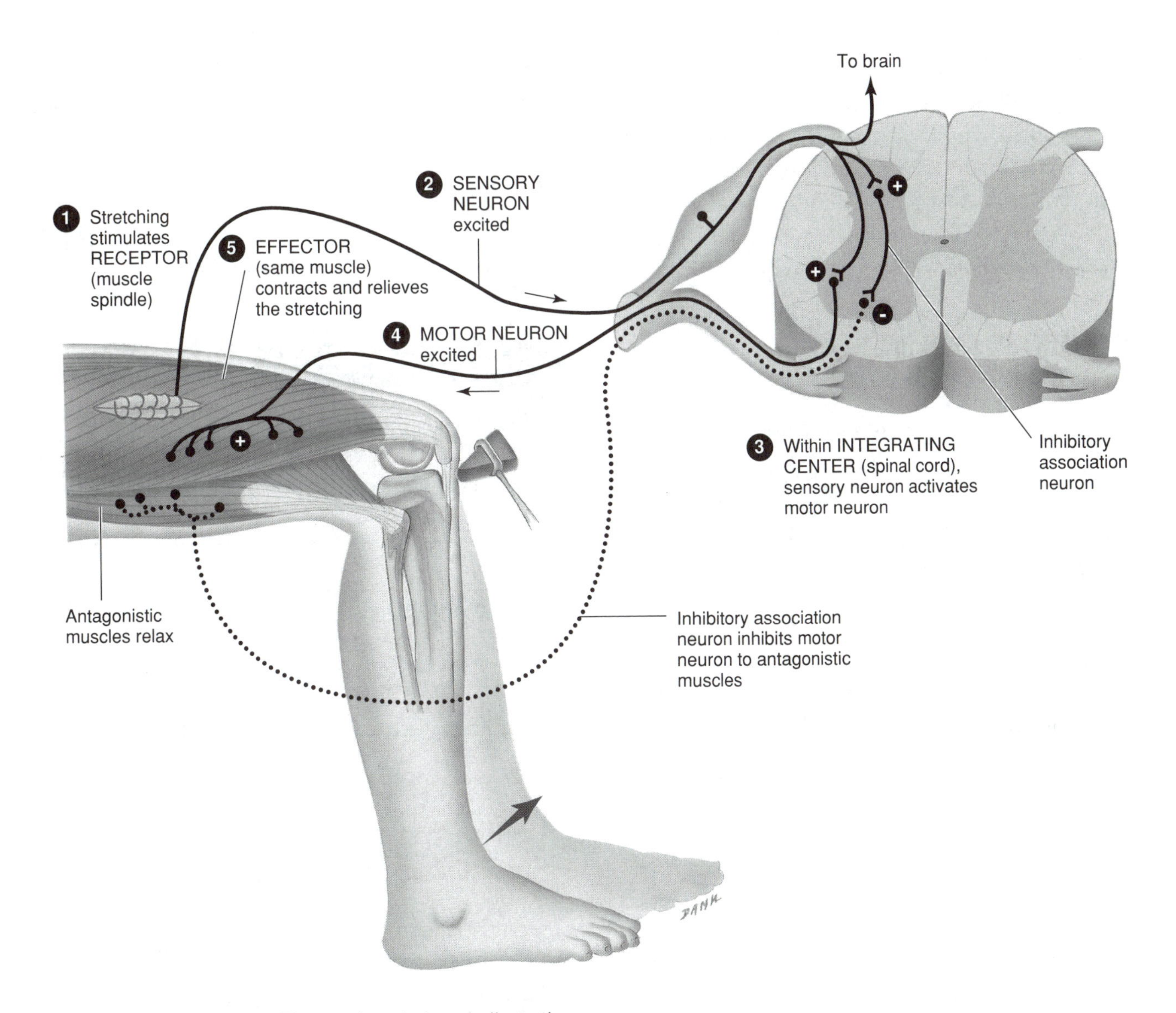

Figure 8.9 Patellar reflex. The numbered steps indicate the pathway of the patellar reflex.

ACTIVITY: Observing the Patellar Reflex

1. Subject is seated on table, leg hanging free and relaxed.
2. Use the reflex hammer to strike the tendon below the kneecap and observe. ■

Ankle Jerk Reflex

This is similar to the patellar reflex and causes the foot to flex in walking. Stimulating the stretch receptor causes the contraction of the calf muscle, which flexes the foot.

ACTIVITY: Observing the Ankle Jerk Reflex

1. Subject is seated on the table, leg hanging free and relaxed.
2. Use the reflex hammer to strike the Achilles tendon, which anchors the calf muscle to the heel bone. ■

Babinski Reflex

This test establishes that the myelin sheath of the spinal cord is present and not damaged. At birth the spinal cord is not fully myelinated. This is the reason why an infant cannot hold its head up, sit up, stand, or walk. As development proceeds and myelination is completed, these events will appear in sequential order.

ACTIVITY: Observing the Babinski Reflex

1. Subject is lying down on the lab table.
2. Using the metal end of the reflex hammer, firmly stroke the arch of the foot starting at the lateral heel, proceed along the lateral side of the foot, and cross over toward the great toe.
3. Record whether the toes fan out or contract under. The normal adult exhibits a negative Babinski reflex. The normal infant exhibits a positive Babinski reflex.

   Since the subject is a mature adult, what would you expect to occur with an infant?

   ____

   ____

   What does this tell you about the direction of myelination?

   ____

   ____

   Can you force an infant to walk at a younger age?

   ____

   ____

   Do all infants develop at the same rate?

   ____

   ____

   What evidence would you offer for your answer?

   ____

   ____

   Alcoholics with polyneuritis and people with multiple sclerosis exhibit a positive Babinski reflex. What type of damage has occurred to the spinal cord?

   ____

   ____ ■

Study Questions

1. Name and describe the three structural types of neurons. Indicate the major type of function of each type.

   ____

   ____

   ____

   ____

   ____

   ____

   ____

   ____

   ____

2. Identify the organs forming the central nervous system.

   ____

   ____

3. Identify the structures forming the peripheral nervous system.

   ____

   ____

   ____

4. Sketch a multipolar neuron and label the major structures.

5. What is a node of Ranvier?

____

____

____

6. Name the structure essential for axon regeneration.

____

7. Identify the meninges. What are their collective functions?

____

____

____

8. What is cerebrospinal fluid? How is it formed and where is it found? What are its functions?

____

____

____

____

____

____

9. Which spinal root has the cell bodies of the sensory neurons?

____

____

____

10. A spinal tap is performed at the level of which vertebrae? Why? (Use your text.)

____

____

____

____

11. Name the three connective tissue wrappings of the nerve and indicate what they wrap.

____

____

____

____

____

____

12. Name the spinal plexuses and the major nerves emerging from each plexus.

____

____

____

____

13. List the structural components of the reflex arc.

____

____

____

____

EXERCISE

Nervous System II: *Brain, Cranial Nerves, and Autonomic Nervous System*

OBJECTIVES

After completing this exercise, you should be able to:

1. Describe and identify the major anatomical regions and structures of the mammalian brain.
2. Explain the function of each structure identified.
3. Name the cranial nerves.
4. For each cranial nerve, cite a major function and a reflex test for that nerve function.
5. Describe what the autonomic nervous system is and what its two divisions (sympathetic and parasympathetic) are responsible for.
6. Identify the major features of the parasympathetic and sympathetic divisions of the ANS.

MATERIALS

- ❑ Plastic model of brain with key to numbered cranial nerves

Materials for evaluating cranial nerve function:

- ❑ Model of the human brain and the key to the model
- ❑ Whole sheep brain
- ❑ Preserved human brain with meninges
- ❑ Human brain (plastic mount)
- ❑ Frontal section of brain through the cerebrum
- ❑ Sheep brains for identifying cranial nerves
- ❑ Cotton swabs
- ❑ Lemon juice
- ❑ Oil of clove and wintergreen
- ❑ Quinine solutions
- ❑ Penlight
- ❑ Tuning fork
- ❑ Safety pins or straight pins
- ❑ All appropriate charts
- ❑ Cotton balls
- ❑ Reflex hammer
- ❑ Sugar solution
- ❑ Ammonia
- ❑ Salt solution

The Brain

The human brain has so many functions that it is hard to imagine what life would be like with a less complex brain. The brain thinks, learns and remembers, allowing us to store information, retrieve it, and apply it to new situations. The brain, not the heart, is the center of our emotions. As such, it is responsible for our ability to love, hate, admire, desire, and be angry or happy. The brain receives and interprets all sensory information, and initiates and coordinates all motor activity. It regulates endocrine function by controlling the release of many hormones. Finally, the brain regulates many involuntary physiologic functions such as heart rate, the rate and depth of breathing, fluid and ion balance, and eating.

The brain has many features in common with the spinal cord, including the presence of meninges, cerebrospinal fluid, and white and gray matter. The brain has the same three protective meninges as the spinal cord; the dura mater, arachnoid, and pia mater. Cerebrospinal fluid (CSF) circulates through the subarachnoid space of the meninges and serves as a fluid cushion, providing nutrients and removing waste. CSF also circulates through the spinal canal and the **ventricles** of the brain. There are four fluid-filled ventricles (cavities) in the brain. There are two lateral ventricles located in the right and left cerebral hemispheres. The third ventricle is located in the diencephalon between the thalamus and hypothalamus. The fourth ventricle is located between the medulla and the cerebellum.

The gray matter of the brain is composed of cell bodies and unmyelinated axons, and the white matter is composed of myelinated axons.

To study all of the important structures of this three-dimensional organ, one must carefully observe the anterior and posterior surface structures and then make sagittal and frontal sections through the brain to see structures on the interior of the brain. Before looking at the anatomy of the sheep brain and a model of the human brain, we will review the functions of the major anatomical features.

Organization of the Brain

The **forebrain** includes the cerebrum and the diencephalon. The midbrain includes the corpora quadrigemina. The hindbrain includes the medulla, pons, and cerebellum. This chapter discusses the major structures and functions of the brain, beginning with anterior structures and moving to the posterior structures.

The Cerebrum

The **cerebrum** is the largest, most superior and anterior part of the brain (Figure 9.1). It is divided into two **hemispheres** by the **longitudinal fissure.** The surface, called the cerebral cortex, is highly convoluted gray matter with numerous **sulci** (grooves) and **gyri** (ridges) that increase the surface area. Each hemisphere is subdivided into four **lobes**, the **frontal**, **parietal**, **temporal**, and **occipital** lobes. The left hemisphere regulates the right side of the body and the right hemisphere regulates the left side of the body. Most people are right-handed because they have left-brain dominance! The white matter of the cerebrum forms bundles of axons called **tracts** that travel in three directions within the brain. The largest tract, the **corpus callosum**, connects the two hemispheres of the cerebrum.

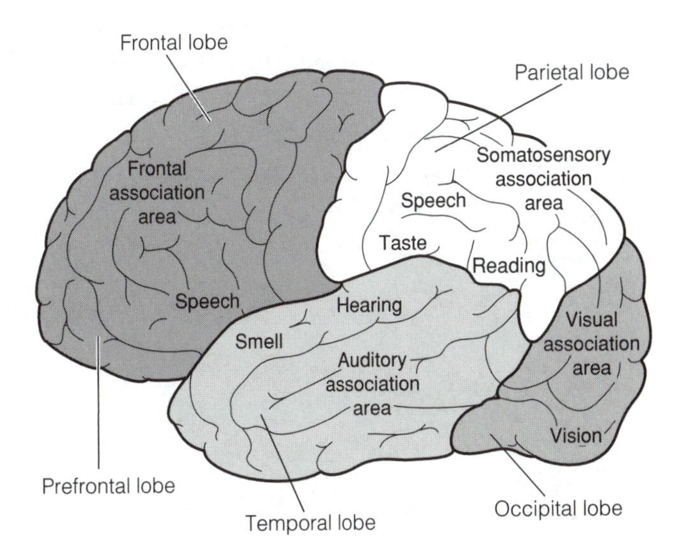

Figure 9.1 Human cerebrum: lobes.

The **olfactory bulbs** receive information relating to the sense of smell. The **prefrontal lobe** is our intellectual center, the center of thought, intelligence, motivation, personality, abstract reasoning, judgement, planning, love, concern for others. It is the most "human" part of the brain. The **frontal lobe** includes the premotor cortex for skilled repetitive activities (like typing) and conditioned reflexes (Pavlov's dog). It is next to the **primary motor cortex**, which initiates voluntary motor activity in skeletal muscles of the legs, arms, torso, and face. Located at the base of the premotor area in the dominant hemisphere only is **Broca's area**, which, as the primary speech center, controls the muscles of speech.

The **parietal lobes** house the **sensory cortex**, which receives transmissions from sensory receptors in the skin and proprioceptors in the muscles and joints. It perceives all sensations except the special senses (taste, smell, sight, and hearing).

Within the **temporal lobe** is the **auditory cortex**, which receives all sound transmission from the ear. You hear with this part of the brain. Also present in the temporal lobes is the **olfactory cortex** for the sense of smell and **Wernicke's area** for comprehension of written and spoken language. When this area is damaged, aphasia results.

The **occipital lobe** houses the **visual cortex**, which receives transmissions from the eye and forms the visual image.

Underlying the cerebral cortex are many different **association areas** that interpret and appreciate the sensations of the primary sensory areas. At the juncture of the parietal, temporal, and occipital lobes is the **gnostic area**, which receives impulses from all sensory association areas and acts as your sensory memory.

Lying under the cerebral cortex and the association areas is the **corpus callosum**, the commissure connecting the right and left cerebral hemispheres. Below the corpus callosum in each hemisphere are the two **lateral ventricles**, which secrete CSF and send it to the third ventricle of the diencephalon.

Emerging from the inferior surface of the cerebrum are cranial nerves I and II. (All cranial nerves will be identified and functions given after we study the brain.)

The Diencephalon

The **diencephalon** is located above the midbrain and under the cerebrum. It includes the thalamus, hypothalamus, and epithalamus (pineal gland). The thalamus has right and left lobes connected by the massa intermedia. It is a sensory relay station, receiving sensory information for pain, heat, cold, touch, pressure, and proprioception and relaying it to the appropriate region of the sensory cortex.

The **hypothalamus**, a small wedge-shaped area under the thalamus, forms the floor and lateral walls of the third ventricle. It is homeostatic in function and contains numerous control centers, including a center for regulating pH and fluid balance, a thermostat for regulating body temperature, and an appestat to regulate feeding. The hypothalamus regulates the activities of the autonomic nervous system and regulates endocrine function by secreting release and inhibitory hormones, as well as by producing the hormones oxytocin and vasopressin (ADH).

The **epithalamus** has the **pineal gland**, which secretes the hormone melatonin. The pineal gland regulates the onset of sexual maturation and serves as a biological clock.

Inferior to the hypothalamus is the **optic chiasma**, formed by the crossing of the two optic nerves. Also found just inferior to the hypothalamus is the **pituitary gland**, which controls many of the other glands and is often called the "master gland."

The Midbrain

The **midbrain** connects the forebrain and hindbrain, and is superior to the pons. It houses the **corpora quadrigemina**, reflex centers for the movement of the head in response to visual and auditory stimuli. Cranial nerves III and IV emerge from the midbrain. Some sources group the midbrain with the brain stem.

The Hindbrain

The **hindbrain** includes the pons, medulla, and the cerebellum. The **pons** is a bridge between the midbrain, the medulla, and the two centers that modulate the activity of the main breathing centers of the medulla. Cranial nerves V, VI, VII and VIII emerge from the pons.

The **medulla oblongata**, the swollen, bulblike end of the brain, is located immediately superior to the spinal cord at the cranial side of the **foramen magnum**. It is composed of ascending and descending sensory and motor tracts that decuss (cross over to the other side), which explains why the right side of the brain controls the left side of the body. The medulla is the most primitive portion of the brain, with islands of gray matter present as nuclei that regulate simple (yet critical) homeostatic functions. In fact, the medulla may keep the body of a comatose person functioning for years.

The autonomic centers present in the medulla include the **cardiac center**, which regulates the rate and force of the heartbeat; the **vasomotor center**, which regulates blood pressure by constriction and dilation of arterioles; and the **respiratory center**, which regulates breathing. The medulla also has reflex centers for vomiting, sneezing, coughing, and swallowing. Cranial nerves IX, X, XI, and XII emerge from the medulla.

Located between the medulla and the cerebellum is the fourth ventricle, a hollow area where cerebrospinal fluid circulates.

The **cerebellum** lies over the medulla and has three lobes, a right and left hemisphere, and the **vermis** between. The cerebellar cortex is highly convoluted and largely gray matter. The **arbor vitae** is the inner white matter, which looks like the branching of a tree. The function of the cerebellum is to integrate and coordinate gross muscle movement. It is a reflex center for posture, equilibrium, and muscle tone.

The **reticular activating system** is composed of small islands of gray matter scattered throughout the midbrain, medulla, and pons. They receive and transmit a steady stream of impulses to the entire cortex. They serve to arouse and alert the cerebral cortex, keeping it conscious.

The **limbic system** includes a number of cortical structures that form a curving border around the corpus callosum. It includes the **cingulate gyrus** (gyrus above the corpus callosum) and the **hippocampus**, a gyrus below the lateral ventricles. As our **emotional** brain, the limbic system lets us experience anger, pleasure, and sorrow. It interacts with other cortical areas that modulate the activity of the limbic system. It plays a major role in **memory**.

ACTIVITY: Studying the Major Parts of the Brain

Identify the parts of the brain using the illustrations provided. Refer to both the sheep brain and the model of the human brain to identify the structures listed.

1. If a sheep brain or human brain is available, observe it within the cranial bones or with the meninges intact.
2. Refer to the whole sheep brain and the model of the whole human brain dorsal view as well as Figures 9.1, 9.2a–c, and 9.3.

   Identify:

- Cerebral hemispheres
- Lobes sulci and gyri
- Longitudinal fissure
- Cerebellum
- Midbrain
- Spinal cord

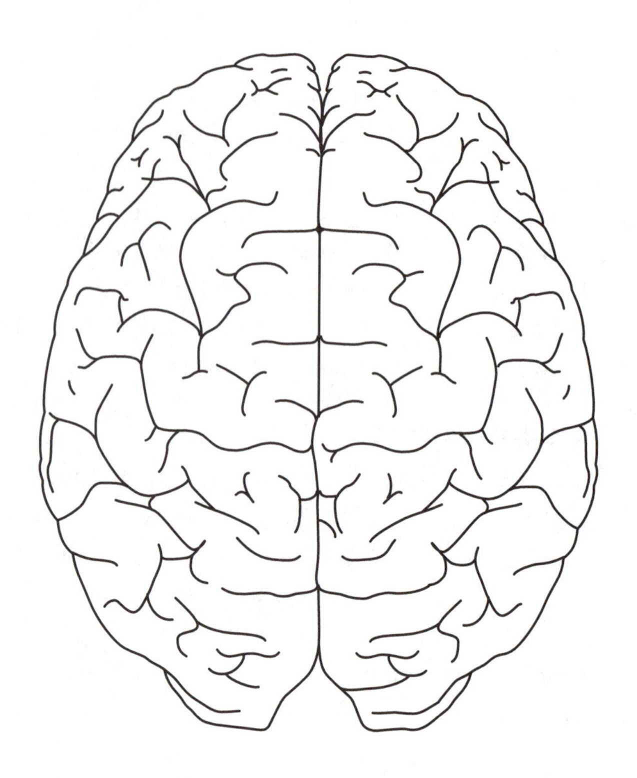

Figure 9.2a Human brain: dorsal view.

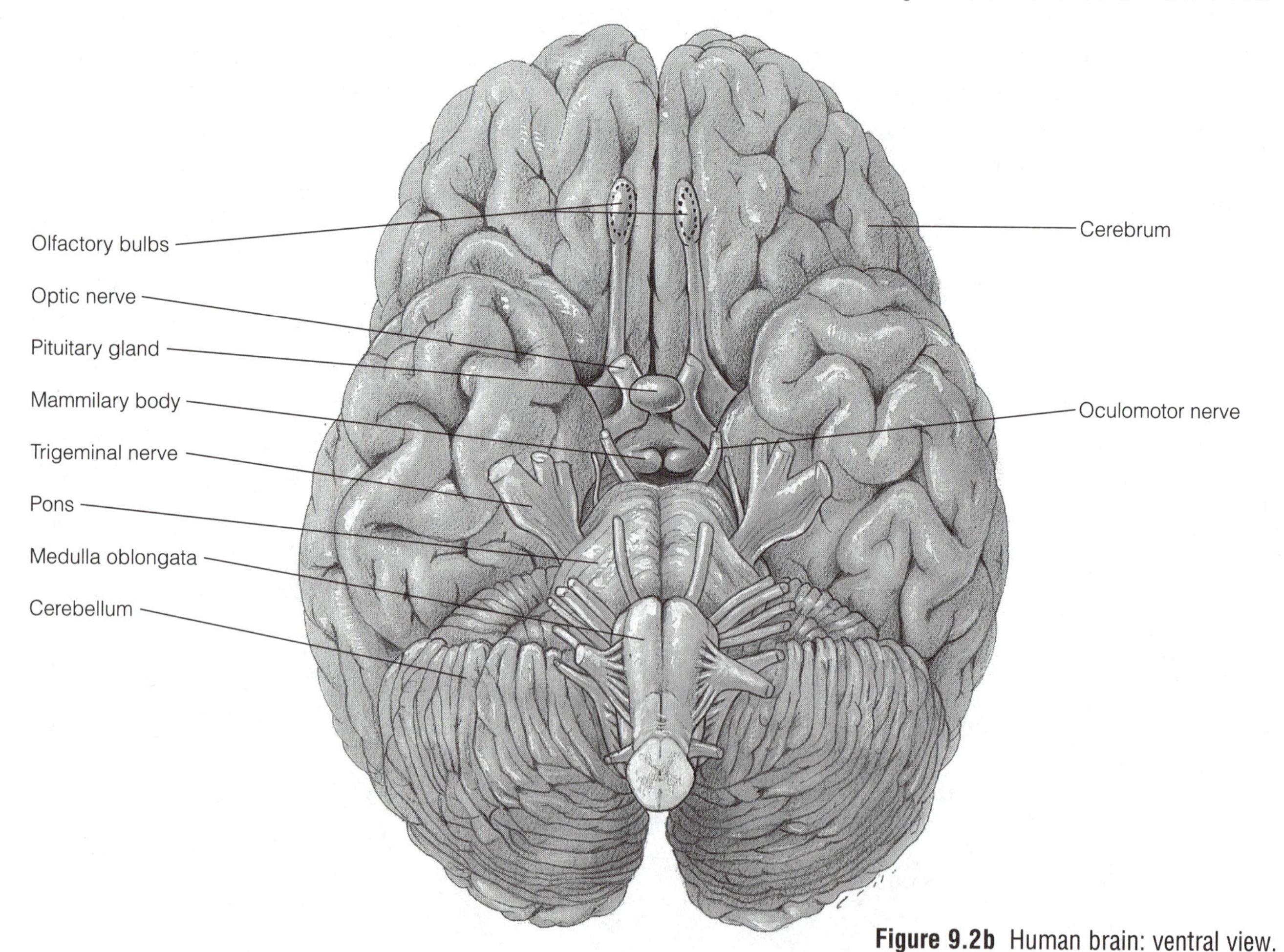

Figure 9.2b Human brain: ventral view.

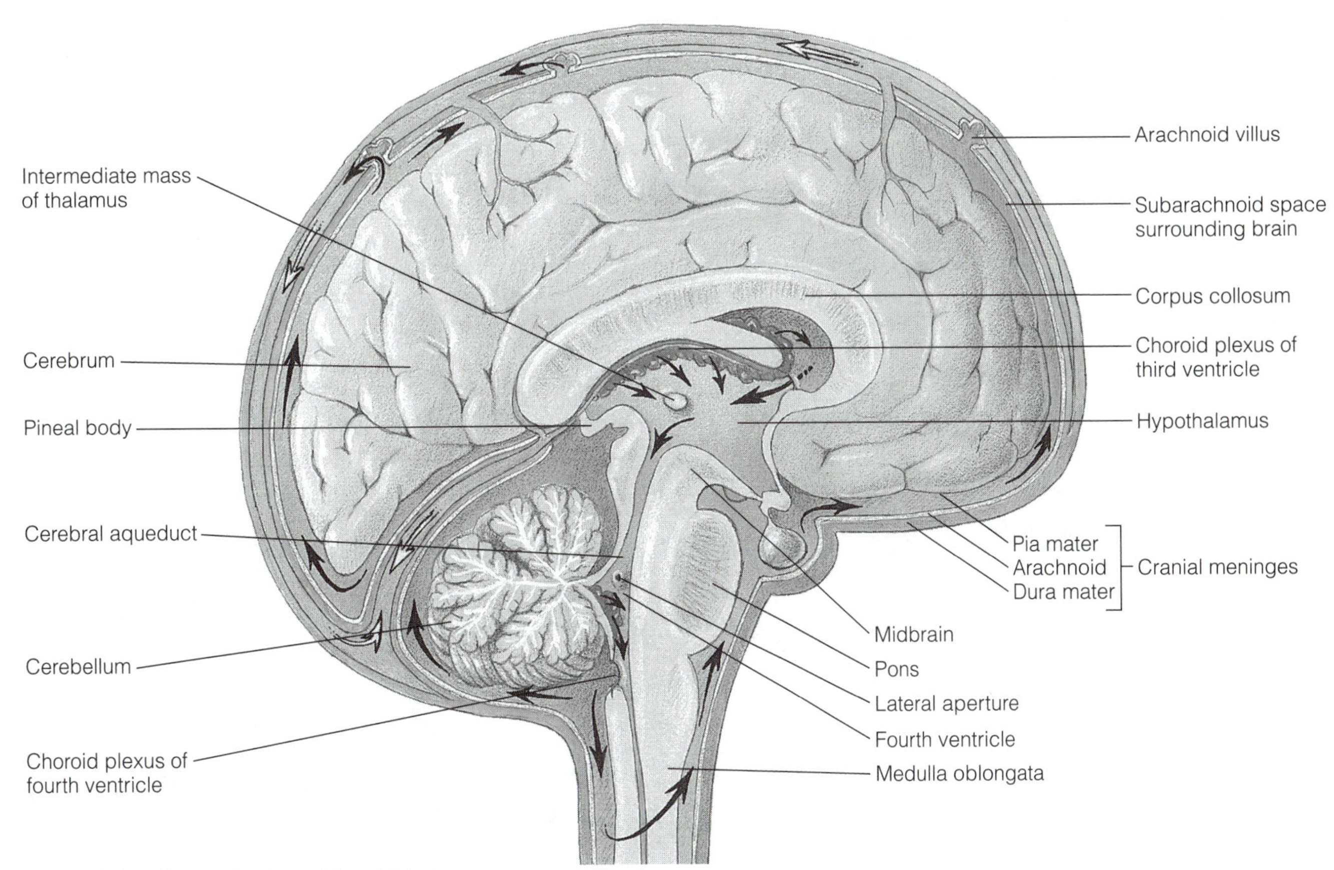

Figure 9.2c Human brain: midsagittal view.

Bend the brain between the cerebrum and the cerebellum to see the pineal gland and the midbrain structures (corpora quadrigemina).

3. Refer to the whole sheep brain and the whole human brain model ventral view as well as Figures 9.2c, 9.3, and 9.5.

   Identify:

- Cerebral hemispheres
- Olfactory bulbs and tract
- Optic chiasma
- Pituitary gland and infundibulum (stalk to which the pituitary gland is attached)
- Pons
- Medulla
- Cerebellum (the sheep brain has one lobe; the human brain has three lobes)
- Spinal cord
- Any intact cranial nerves (Figure 9.5 on page 79)

4. Brain dissection: insert a very sharp, long-bladed knife into the longitudinal fissure and make a sagittal section through the sheep brain.

5. Refer to the cut surface of the sheep brain and the sagittal section of the human brain model as well as Figure 9.4.

   Identify:

- Cerebral hemispheres
- Corpus callosum
- Lateral ventricle
- Optic chiasma
- Pituitary gland
- Thalamus
- Hypothalamus
- Pons
- Medulla
- Cerebellum and arbor vitae
- Fourth ventricle
- Spinal cord

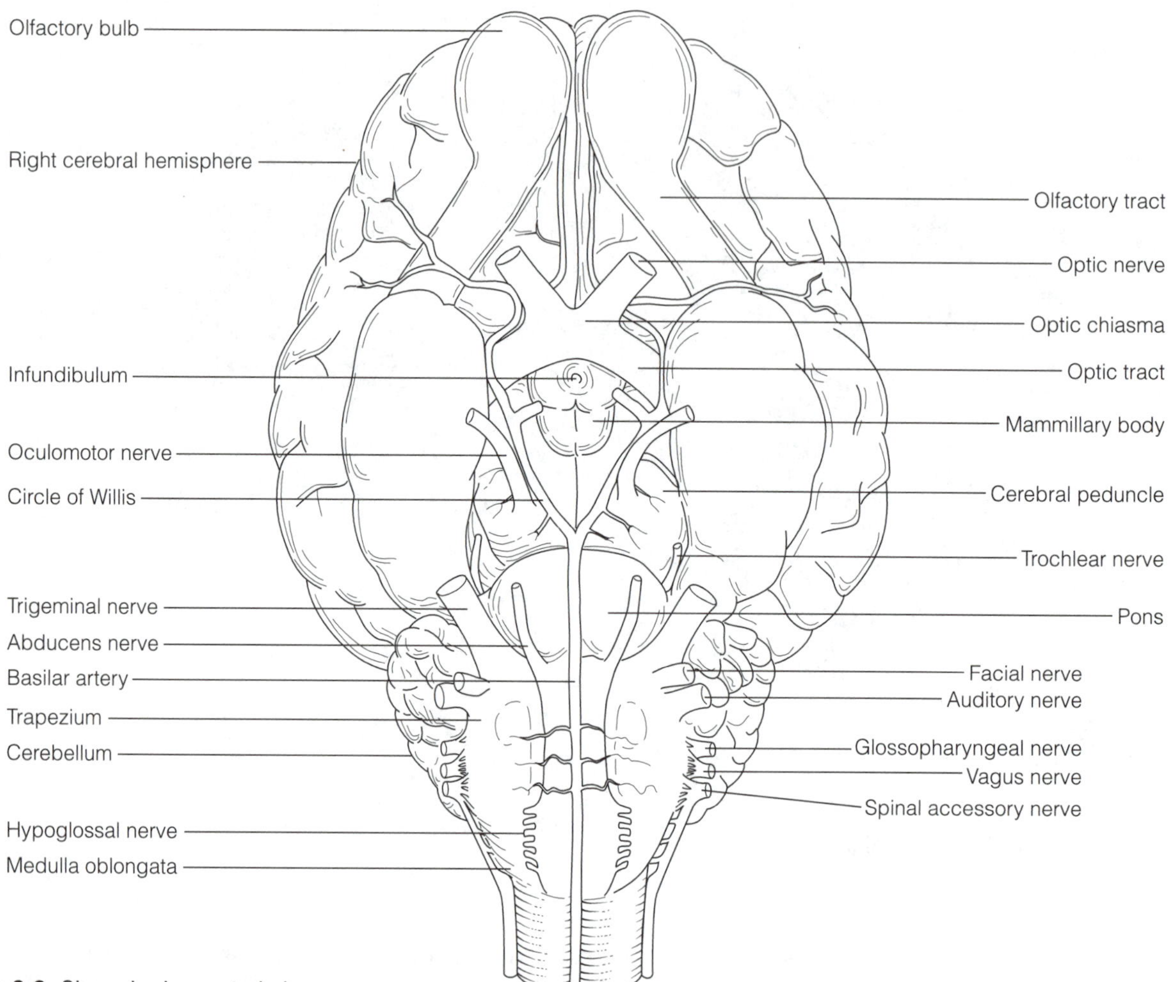

Figure 9.3 Sheep brain: ventral view.

6. Refer to the demonstration frontal cut of the sheep brain as well as Figure 9.4.

   Identify:

- Cerebrum
- Cerebral cortex
- White matter
- Corpus callosum
- Lateral ventricles
- Third ventricle
- Thalamus
- Hypothalamus
- Cranial nerves ■

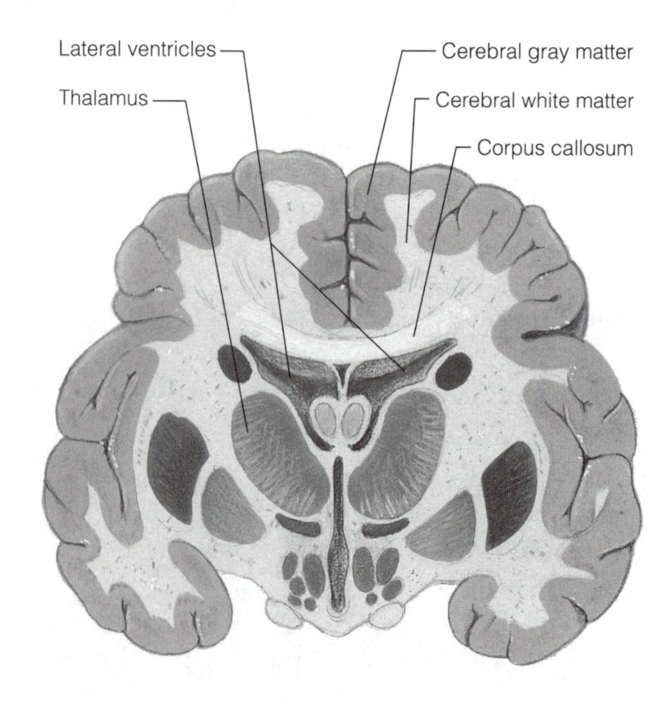

Figure 9.4 Human brain: frontal section.

Cranial Nerves

There are twelve pairs of cranial nerves (Figure 9.5). While all are mixed nerves, some are considered primarily sensory, or primarily motor. All innervate the head and neck *except* for one that descends into the thoracic and abdominopelvic cavities.

ACTIVITY: Studying the Function of the Cranial Nerves

Working with a partner, test the function of the cranial nerves as indicated in Table 9.1. (Instructor's option.)

Was the subject responsive to all of the above tests? If not, indicate which he or she couldn't carry out and what the implication might be.

____

____

____

____

____

Which nerve descends into the abdominopelvic and thoracic cavities?

____

____

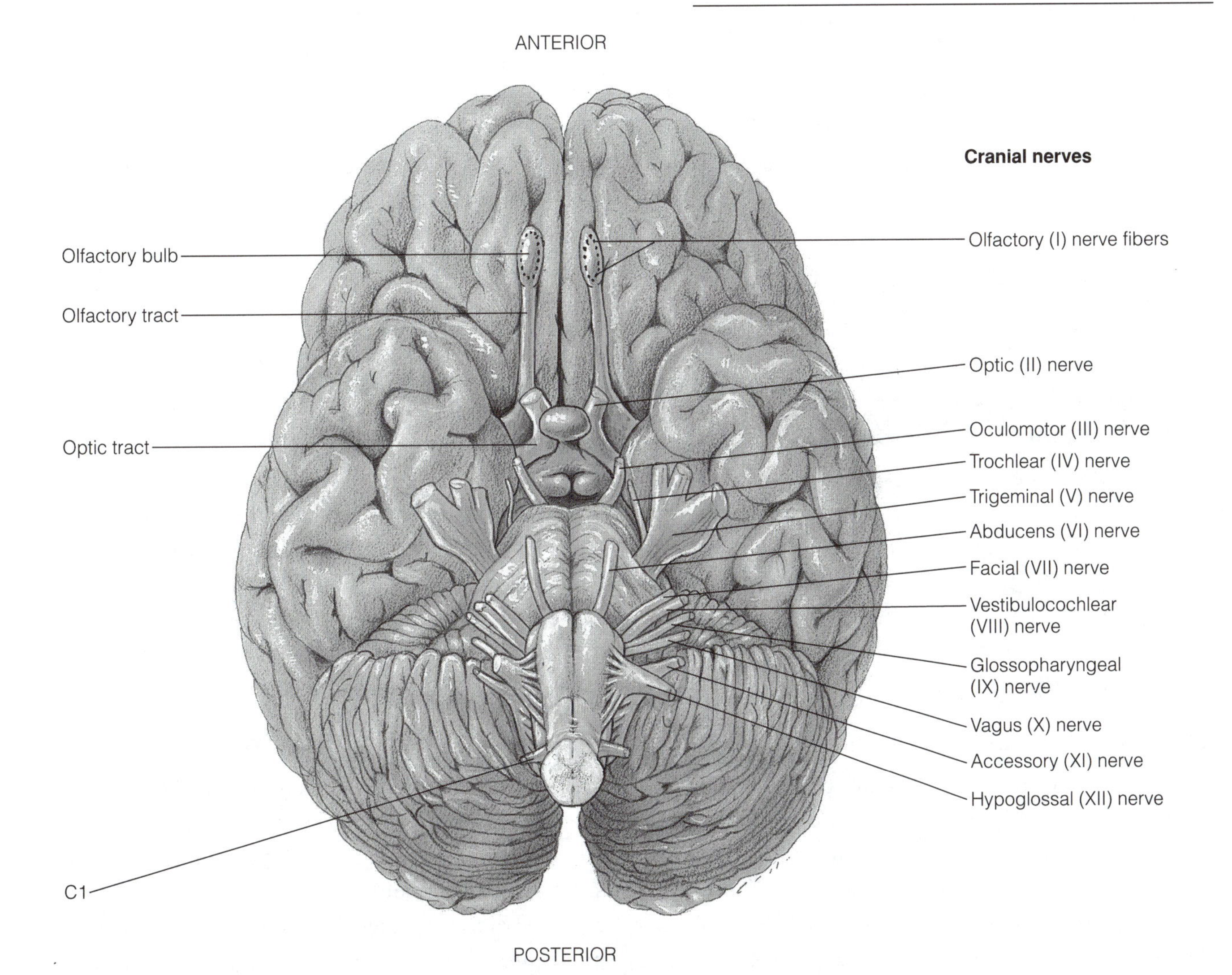

Figure 9.5 Human brain and cranial nerves: ventral view.

Table 9.1 Testing the Function of Cranial Nerves

Cranial Nerve	Function	Function Test
I. Olfactory	Smell	Sniff to identify oil of cloves, wintergreen
II. Optic	Vision	Ability to read this manual
III. Oculomotor	Eye movement, focusing, regulating amount of light	Pupillary reflex; eye movement up, down, right and left
IV. Trochlear	Eye movement	Tested with nerve III
V. Trigeminal	Sensation in the eye, nose, and mouth; chewing	Response to a pinprick to the skin near the eye, nose, and mouth; ability to chew
VI. Abducens	Eye movement	Tested with nerve III
VII. Facial	Facial expression, tears and saliva flow	Test anterior tongue for sweet, salty, and sour; test tear flow with NH_3 (sniff cautiously)
VIII. Auditory	Hearing and equilibrium	Test hearing with a tuning fork; test equilibrium—stand on one foot, with eyes closed
IX. Glossopharyngeal	Saliva flow, taste, muscles of swallowing and speech	Gag reflex, uvula straight, speak, cough, bitter taste (posterior of tongue)
X. Vagus	Parasympathetic control of organs in thoracic and abdominal cavities	Difficult to test in class
XI. Spinal accessory	Muscles of neck, head and throat, swallowing	Turn head right, left, up and down; swallow
XII. Hypoglossal	Tongue	Protrude and retract tongue

Autonomic Nervous System

This division of the nervous system is responsible for the normal calm functioning of the body and for mobilizing the responses of the body to stressful or dangerous situations (Figure 9.6). Its sympathetic and parasympathetic divisions provide this dual function. The activitiy of this division is entirely involuntary and consists of motor nerves only.

Sympathetic Division

This is the "fight or flight" division of the ANS. The sympathetic nerves emerge from the thoracic and lumbar segments of the spinal cord and innervates many organs. This division releases the neurotransmitters **epinephrine** and **norepinephrine**. Some of the diverse effects of these hormones include dilation of the pupils in order to improve vision, inhibition of digestive function, increased rate and force of the heartbeat, relaxation of the bronchi to facilitate breathing, vasoconstriction of most blood vessels to raise blood pressure, vasodilation of blood vessels in the legs to increase blood flow, and release of glucose and fatty acids from storage to the bloodstream. All of these responses leave your body better prepared to respond to a physiologically stressful event.

Parasympathetic Division

The parasympathetic division is considered by some to be the "rest and repose" or "normal function" division. It keeps your body "humming" along under normal nonstressful conditions. The parasympathetic nerves emerge from the brain and sacral region of the spinal cord. This division releases the neurotransmit-

(text continued on page 82)

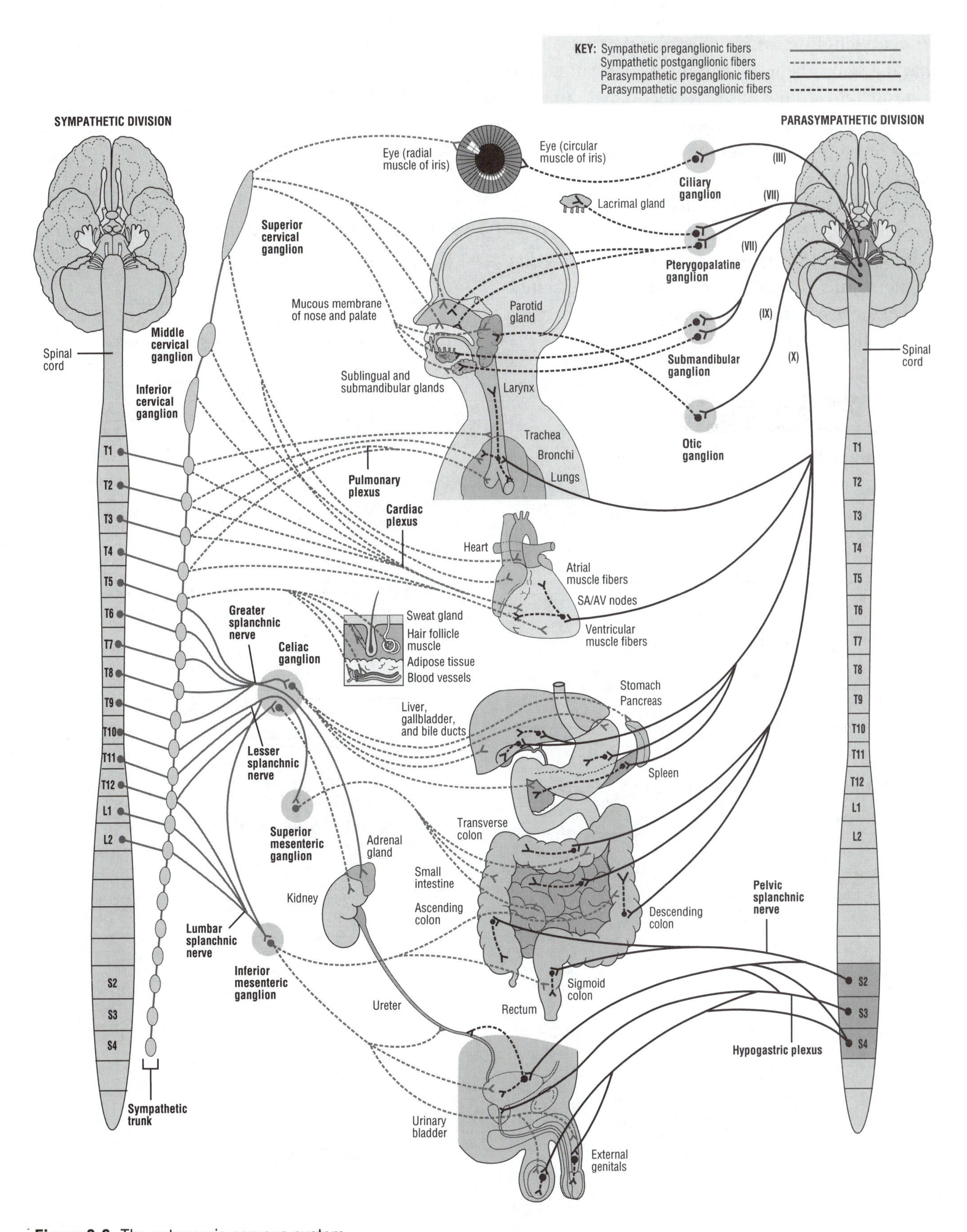

Figure 9.6 The autonomic nervous system.

ter **acetylcholine**, which innervates the same organs that the sympathetic division innervates, but has the opposite effect. It constricts the pupils and the bronchi, promotes vasodilation and the reduction of blood pressure, and reduces the rate and force of the heartbeat. The parasympathetic branch stimulates the flow of saliva and digestive juices and digestive contractions. Figure 9.6 is a classic illustration of the nerves emerging from the two divisions and traveling to the organs that they innervate. This illustration conveys considerable additional information about the autonomic nervous system.

Study Questions

1. List three functions of the cerebrum.

____

____

____

2. What is the function of the cerebellum?

____

____

3. List the functions of the medulla.

____

____

____

4. List the functions of the hypothalamus.

____

____

____

5. Identify the cranial nerves that are involved in eye movements.

____

____

____

____

6. Identify the cranial nerve which is involved in equilibrium.

____

____

7. This cranial nerve regulates head movement:

____

____

8. Which division of the ANS would be involved in the beating of your heart when you are calm and relaxed?

____

____

9. Identify some sympathetic responses to stress and indicate how they prepare you for coping with the stressful situation.

____

____

____

____

10. Why is exercise beneficial to someone with asthma, whose bronchi are constricted?

____

____

____

____

11. Why might stress raise blood pressure?

____

____

____

____

12. Why shouldn't an athlete eat a large meal prior to participating in an athletic activity?

____

____

____

____

13. Is the stress response appropriate to the type of stressful situations one encounters in modern life? If so, why?

____

____

____

____

If not, why?

____

____

____

____

EXERCISE

10 The Senses

OBJECTIVES	MATERIALS
After completing this exercise, you should be able to: 1. Understand the concept of sense reception and identify the major sense receptors of the human body. 2. Identify the main structures of the eye and ear in the sheep eye and human eye and ear models. 3. List and describe the basic functions of the structures found in the eye and ear. 4. Describe the basis of the eye and ear tests performed in the lab, and the results of these tests.	❑ Human torso models ❑ Preserved sheep eyes (one eye per two students) ❑ Human eye and ear models ❑ Tuning forks ❑ Snellen eye chart ❑ Astigmatism eye chart ❑ Color blindness test kit ❑ Two-point discrimination kit ❑ Test tubes and pencils ❑ Sugar and salt ❑ Paper cups ❑ Pencils and rubber band ❑ Swabs

General Sensation

Chains of neurons connected to muscles or glands allow us to react to the environment with thought, movement, or some other action. Neurons along these pathways communicate through chemicals called **neurotransmitters**. However, in order to pick up a stimulus from the environment, there must be a special class of neurons that can be stimulated by environmental factors instead of neurotransmitters. The specialized neurons that can be stimulated by environmental factors are called **receptors**. These sensory receptors are usually categorized by the kind of stimulus to which they are sensitive.

Chemoreceptors

Chemoreceptors are stimulated by chemicals that "plug into" receptors found at the end of the sensory neuron. The chemical and receptor fit together like a lock and key. Chemoreceptors in the tongue and nose are responsible for our senses of taste and smell.

Pain Receptors

Pain receptors may actually be a kind of chemoreceptor sensitive to chemicals produced when nearby cells are damaged or destroyed. Although most prevalent in the skin, pain receptors are also found internally.

Mechanoreceptors

Mechanoreceptors are sensitive to mechanical or physical changes around them. Often they are coated with layers that will help to trigger an impulse when their shape changes.

Following are some other mechanoreceptors and their functions:

- **Stretch receptors** such as those in the lungs prevent overexpansion.
- **Proprioceptors** detect stress placed on joints and muscles.
- **Hair cells** of the ear respond to mechanical disturbance translated into nervous impulses which are eventually interpreted as sound.
- **Baroreceptors** sense changes in blood pressure and are a very delicate form of mechanoreceptor.
- **Tactile receptors** sense "touch" or contact with the skin.

Which tactile receptors did you learn about when you studied the skin? What were their functions?

____

____

Photoreceptors

Photoreceptors, as the name implies, are stimulated by light and are found exclusively in the retina of the eye.

Thermoreceptors

Thermoreceptors respond to temperature changes. Free nerve endings in the skin seem to be responsible for this sensation, although the precise mode of action is not clearly understood.

Receptor Adaptation

When exposed to the same degree of stimulus over time, many receptors cease responding to the stimulus. The receptors are said to have adapted to the stimulus. One of the reasons we do not feel the clothes on our bodies, except for when we put them on and take them off, is that the receptors stop responding to the mild tactile stimulus after a few minutes. Adaptation does not just apply to tactile receptors. Olfactory chemoreceptors also adapt, as evidenced by the fact that we tend to notice the way a room smells only when we first enter it. To a lesser extent, thermoreceptors also tend to fire less when exposed to the same temperature, and respond best to changes in temperature. This phenomenon also explains why most receptor categories respond best to stimulus changes rather than repeated exposure to the same level of stimulus.

ACTIVITY: Studying Receptor Function

Chemoreceptor Function

1. Dip a moistened swab into sugar and touch the tip of your tongue with the swab. (Your tongue must be moist, not dry.) Do you taste the sweetness?

   ____

2. Rinse your mouth thoroughly and place the same amount of sugar near the back of your tongue. Repeat again on the sides of the tongue. Do you taste the sweetness in these areas?

   ____

3. Repeat the above procedure for salt. Where do you sense the strongest salty taste?

   ____

4. Use your textbook to look up the location of the various taste receptors of the tongue and list them below.

   ____

   ____

   ____

   ____

   ____

5. Are taste chemoreceptors specific to categories of chemicals? __________. Explain your answer and your results.

   ____

   ____

   ____

   ____

   ____

Discrimination Test

1. Work in pairs. Obtain the two-point discrimination kit set out in the lab.
2. The "subject" partner closes his or her eyes while the "tester" partner touches the subject's fingertips with testing blocks. The tester randomly alternates use of differently spaced blocks.
3. Record the smallest separation that could still be identified as two points. ____________________.
4. Repeat for the palm of the hand and record the smallest separation that could still be identified as two points. ____________________.
5. Repeat for the anterior forearm and record.

   ____________________.
6. Repeat for the back of the neck and record.

   ____

   Place these areas in order from most sensitive to least sensitive.

   ____

   ____

   Propose an explanation for the pattern of tactile receptor density you obtained for the human body.

   ____

   ____

   ____

   ____

Proprioceptor Function

1. Work in pairs. Obtain three paper cups and add water to two of the cups until one is about half full and the other is about one-fourth full.
2. The subject partner closes his or her eyes, and the tester partner hands the subject the two cups. Can the subject tell which cup is heavier? __________.
3. Using a third cup of water, add a small amount of water (about 1/4 cup) to either cup (the subject's eyes should still be closed). Can the subject tell to which cup water was added? ____________________.

   Explain how the subject's receptors were working during this experiment.

   ____

   ____

   ____

   ____________________ ■

The Human Eye

The human eye (Figure 10.1) could be compared to a camera, with the retina a reusable layer of photographic film. Like the camera, the eye (with eyelid open) focuses a light image on its posterior area. Every 1/10 of a second, the image is cleared and a new image is interpreted by the brain. In the next section, we will examine the parts of the eye involved in producing this image.

The three tissue layers of the eye are the outer sclera, the middle choroid, and the inner retina. All three layers are adapted in interesting ways for the various roles they play in human vision.

Sclera

The **sclera** is the outer layer of tough, white connective tissue that is modified on its anterior side to allow light to pass through. This small transparent area on the anterior portion of the sclera is called the **cornea**.

Choroid

The choroid is the middle pigmented tissue that holds a large portion of the blood supply for the retina. The dark pigmentation of the choroid prevents scattering and reflecting of light that would distort the image. It is interesting to note that some red light is permitted to reflect back from the choroid, causing the red-eye effect in some photographs.

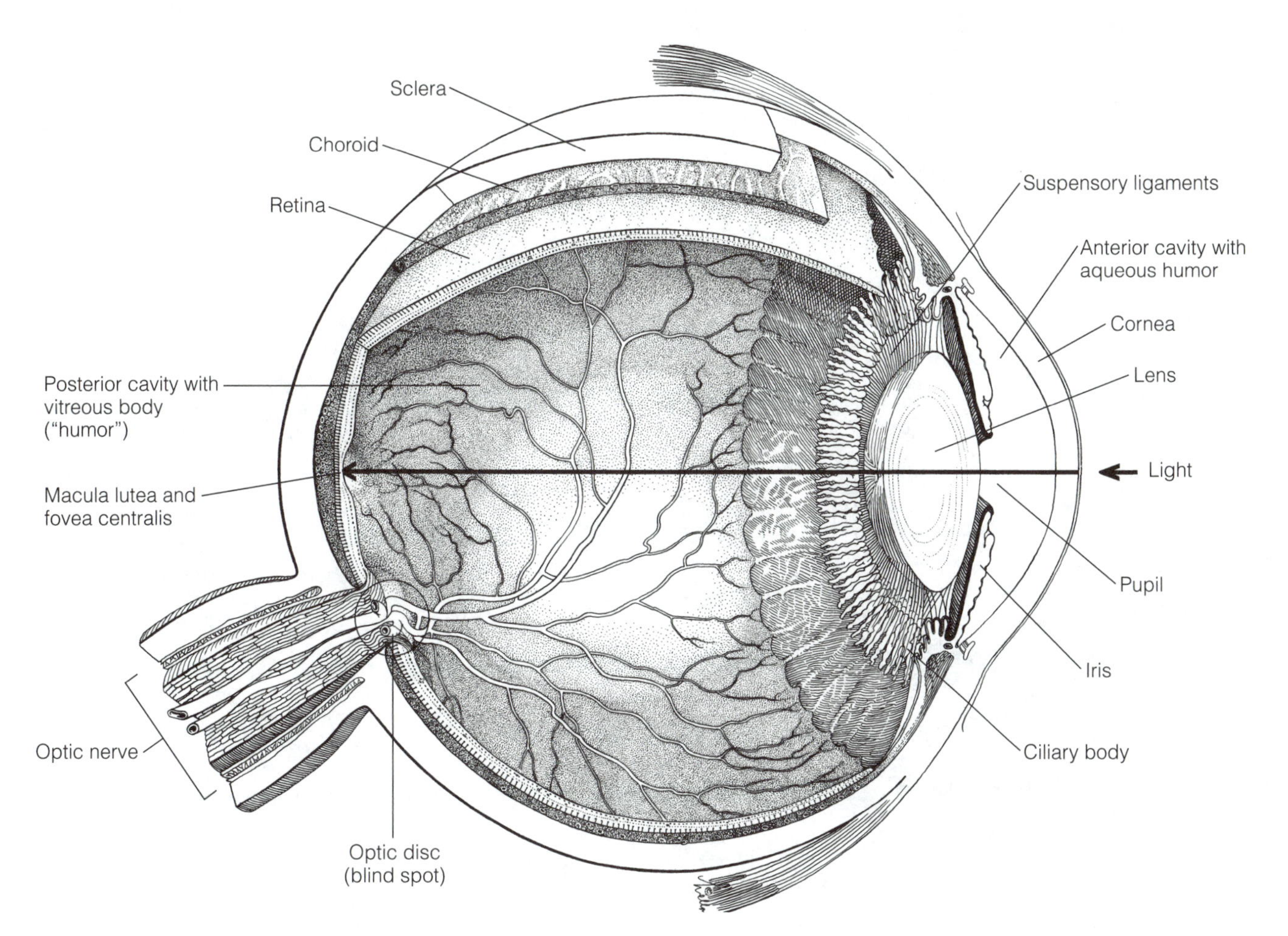

Figure 10.1 The human eye: sagittal section.

The choroid also forms two important muscular structures near the cornea: 1) the **iris**, a ring of smooth muscle that regulates the opening of the pupil; and 2) the **ciliary body**, which changes the tension on (and shape of) the lens. Made of transparent crystalline proteins, the **lens** is held in the center of the ciliary body by tiny **suspensory ligaments** similar to the way a trampoline is held to its frame by numerous slender cords. The lens focuses the image on the retina, but does so upside down. However, the brain reorients the image.

Retina

The retina is the inner nervous tissue layer of photoreceptors, which set off a nerve impulse when struck by light. Two kinds of photoreceptors, rods and cones, are found in the retina. **Rods** are very sensitive and are triggered to produce impulses when struck by very low levels of light, such as occurs at night. The rods come in only one variety, so night vision is black and white. **Cones** come in three varieties (red, green, and blue), allowing the retina to transmit an image in color. These specialized color photoreceptors require much more light in order to work, so they function only in relatively good light conditions. A dense collection of cones called the **macula lutea** are found at the back of the eye near the center of focus, or **fovea centralis**. The image or pattern of light on the retina is transmitted and reassembled into a similar pattern in the brain, which compensates for the upside down orientation of the image.

Between the cornea and the lens in the anterior section is a fluid called **aqueous humor**. The larger posterior section has a thicker fluid called the **vitreous**

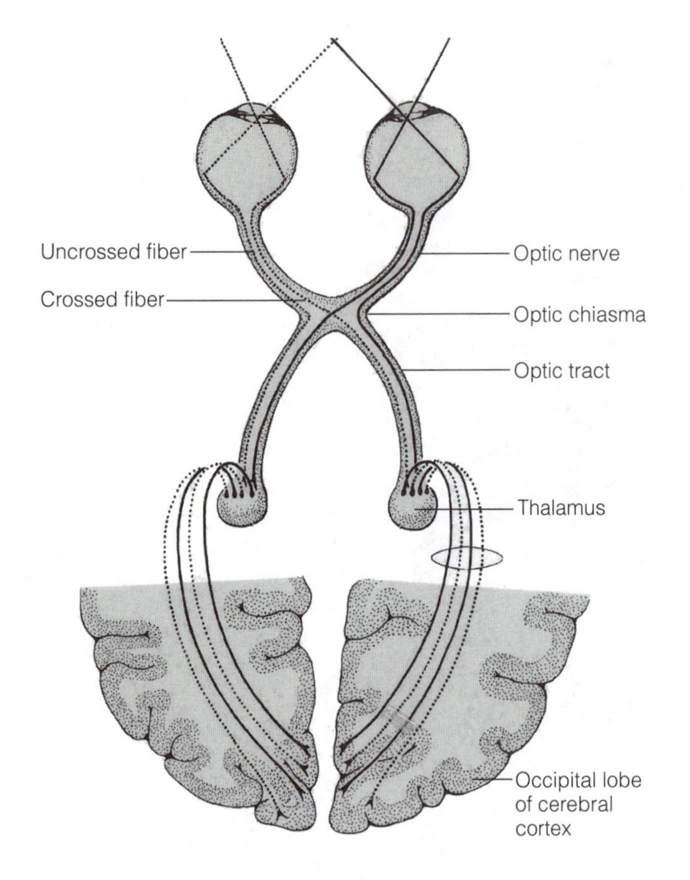

Figure 10.2 The human visual pathway.

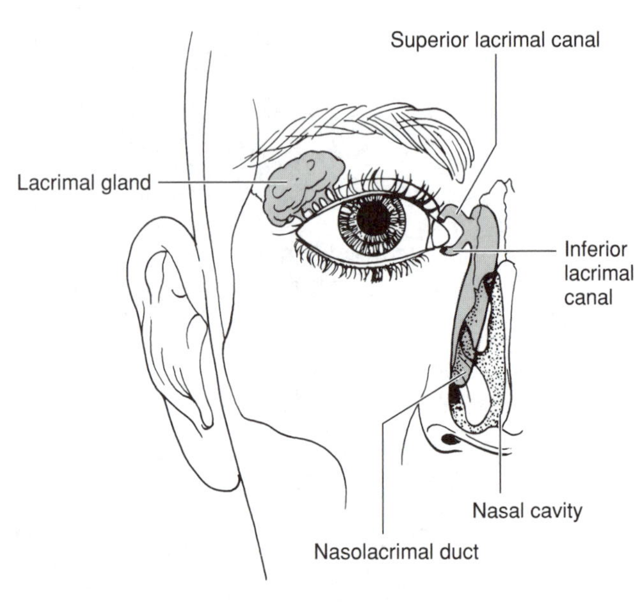

Figure 10.3 Structures that protect the eye.

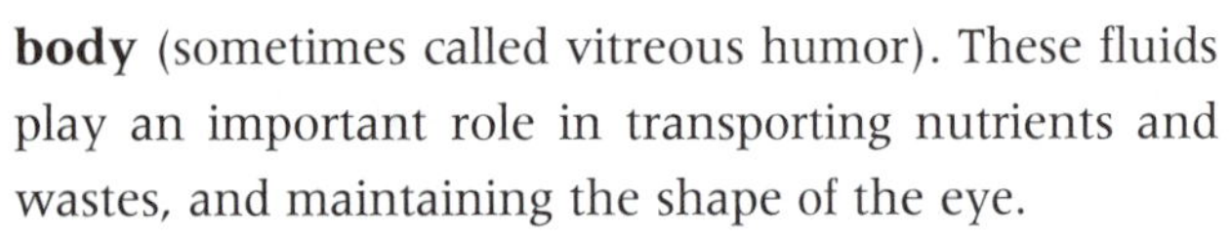

body (sometimes called vitreous humor). These fluids play an important role in transporting nutrients and wastes, and maintaining the shape of the eye.

Light entering the eye passes through the cornea, aqueous humor, pupil, lens, and vitreous body before it strikes the retina. Precise focusing requires thickening or thinning of the lens so that it bends incoming light rays to a greater or lesser degree. The varying tension exerted on the lens by contraction and relaxation of the ciliary body changes its shape and allows for focusing on near or distant objects.

The image now focused on the retina is translated into a pattern of nerve impulses that exit the eye by the optic nerve (Figure 10.2). The exit point of the optic nerve displaces photoreceptors, creating a blind spot called the **optic disc**.

Because it is a very delicate organ, the eye is protected from mechanical damage by the bony socket of the skull and a layer of adipose tissue that cushions the eye socket (Figure 10.3). Blinking and tears from the lacrimal glands help to keep small objects from entering or remaining in the eye.

What structures of the eye translate light energy into nervous impulses during the day?

____

At night? ____________________________________.

What do you think causes color blindness?

____

____

____

ACTIVITY: Studying the Structure and Function of the Eye

Sheep Eye Dissection

1. Carefully make an incision in the sclera with a scalpel about 1/4 inch away from the edge of the cornea.
2. Starting with the hole made by the scalpel, use scissors to cut a circle around the cornea, leaving a 1/4-inch border of white sclera.
3. Remove the cornea and border. This separates the eye into two pieces: a posterior 2/3 and an anterior 1/3.
4. From the inside, slowly remove the lens and observe the suspensory ligaments stretch and break.
5. Note the two round, dark structures—the smoother, more anterior iris and the ridged, more posterior ciliary body.
6. Carefully peel back the cream-colored retina (it may have already detached during the dissection), and observe the dark choroid coat. Note that one portion of the retina stays firmly attached. This is the optic disk, where the optic nerve exits the eye.
7. Use a forceps and pin probe to separate the choroid from the sclera.
8. After viewing the cornea and noting its placement, use scissors to remove the cornea so that the iris and pupil can be observed from the anterior side.
9. Locate the following structures on the sheep eye and the human eye models:

- Cornea
- Pupil
- Ciliary body
- Vitreous body
- Choroid
- Optic nerve
- Iris
- Lens
- Suspensory ligaments
- Retina
- Optic disk
- Sclera

You may notice a lighter blue area on the otherwise dark choroid. This tapetum lucidum allows for the reflection of light back through the retina a second time, enhancing vision in relatively dark conditions. The tapetum lucidum is even more pronounced in nocturnal hunters such as cats and owls, but is absent in humans.

Eye Tests

Visual Acuity

1. Stand 20 feet away from the Snellen chart, cover one eye, and read the letters out loud. Your lab partner will stand next to the chart to verify your reading. The numbers to the left of the last line you read correctly pertain to the vision rating for that eye. If you wear glasses, perform the test both with glasses on and off.
2. Repeat for the other eye. Note that the set of numbers to the side of the row of letters always starts with 20. This number simply corresponds to the number of feet you are standing from the chart. The second number refers to the distance that a person with "normal" vision would be standing from the chart if that was the person's last correct line. For example, if you could only read the big "E" (20/200), that means that a person with normal vision could see that letter from 200 feet away.
3. Record your visual acuity for each eye.

____

____

Astigmatism

All lines should be of equal thickness and darkness when you look at the astigmatism chart (look through one eye at a time). If the lines are unequal, you may have an irregular curvature of the lens or cornea. This condition can be corrected by glasses.

Binocular Vision

1. Have your lab partner hold a test tube 2 to 3 feet away from you.
2. Starting with your arm at your side, quickly but carefully place a pencil inside the test tube, eraser side down. Repeat five times.
3. Next, close one eye and attempt to place the pencil in the test tube. Remember to do this quickly but carefully. Repeat five times for each eye.

4. Record the number of times that you could successfully insert the pencil into the test tube with both eyes open ________________. Left eye only ______________. Right eye only ______________.

   What does the term "20/40 vision" really mean?

   ____

   ____

   ____

   Why is it more difficult to place the pencil in the tube with one eye closed?

   ____

   ____

   ____

Color Blindness

Observe the color blindness tests on display. Does seeing how these tests are designed help you answer the earlier question on the cause of color blindness? If so, revise your explanation regarding the cause of color blindness.

____

____

____ ■

The Ear

Although the outer ear and middle ear are involved in the conduction of sound, the inner ear is both an organ of hearing and of balance. The **outer ear** includes the pinna and acoustic meatus (auditory canal). The **middle ear** is a small cavity in the temporal bone. The boundary between the outer and middle ear is the **tympanic membrane**. Found in this cavity are the small bones, the malleus, incus, and stapes. The **eustachian tube** is a connection between the middle ear and throat. The **inner ear** is made up of the coiled, snail-shaped, tubular **cochlea**, a thick **vestibule**, and three **semicircular canals** (Figure 10.4).

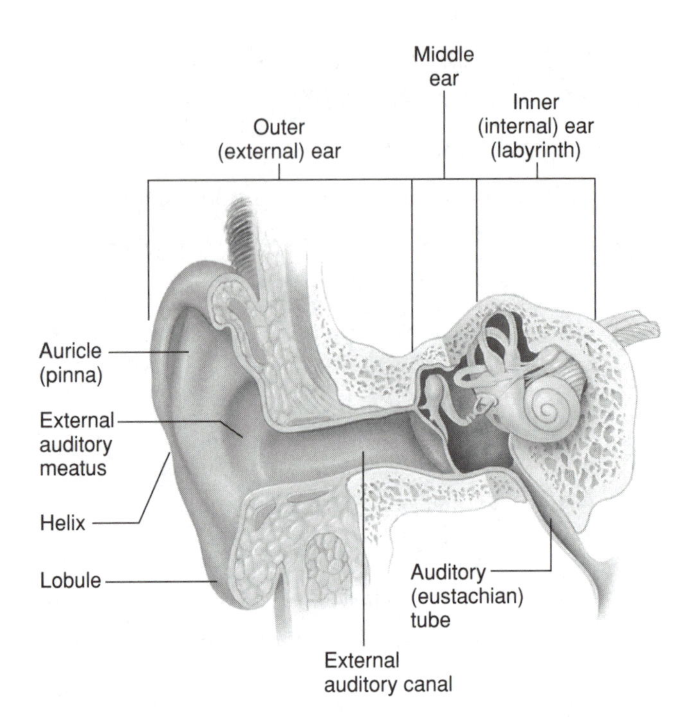

Figure 10.4 Anatomy of the human ear and accessory structures.

Hearing

The pathway between the pinna and the inner ear changes the air disturbances we call sound into motion that will eventually vibrate fluid in the cochlea. The outer ear funnels sound vibrations toward the conduction system of the middle ear. The tympanic membrane (the ear drum) vibrates in response to sound waves and transfers this movement to the malleus, incus, and stapes. The eustachian tube maintains pressure equilibrium in the otherwise sealed-off middle ear by indirectly connecting to the outside via the throat. When the stapes vibrates against an opening in the cochlea, called the **oval window**, the vibrations are converted into fluid movement. Tactile receptors called **hair cells** in the **organ of Corti** are stimulated by this fluid movement.

As you can see in Figure 10.5, the cochlea is a long, coiled tube, and different sound frequencies disturb the fluid in different areas of the tunnel. This tactile reception in different areas of the cochlea is interpreted as the variety of sounds we can hear. Just as you can tell the difference between something touching your wrist and something touching your elbow,

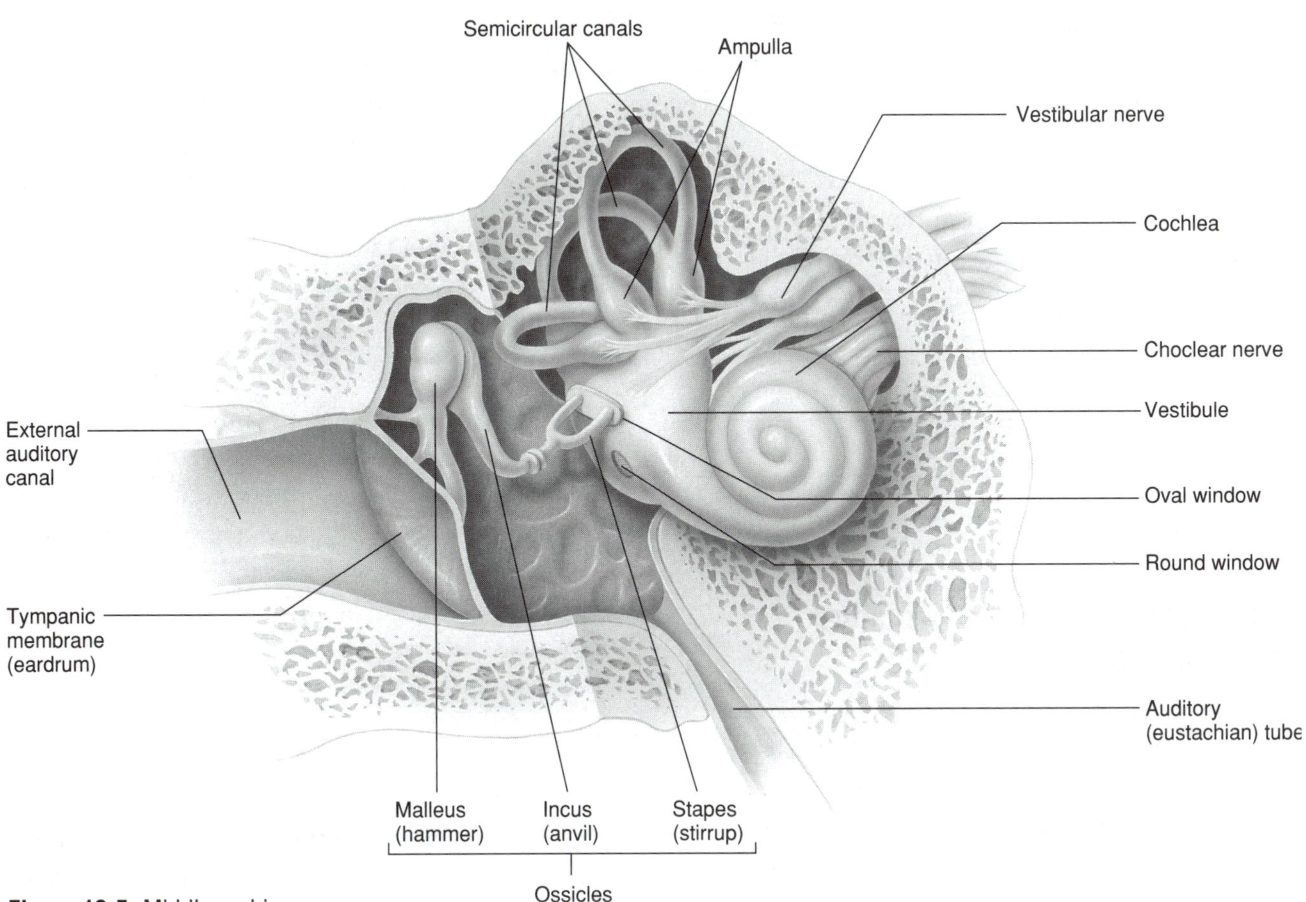

Figure 10.5 Middle and inner ear.

the brain can differentiate high and low frequencies based upon whether the closer or farther areas of the cochlea have their hair cells "touched" by fluid vibration. Once the hair cells are stimulated, the **vestibulocochlear nerve** carries the nervous impulses to the auditory area of the brain.

Name, in order, the structures involved in the conduction of sound from the pinna to the brain.

____

____

____

What kind of receptors are used to produce your sense of hearing?

____

____

____

What "organ" in the inner ear translates vibrations into nervous impulses?

____

____

____

Balance and Equilibrium

The vestibule and semicircular canals of the inner ear also make use of tactile receptors, but for a sense of balance rather than hearing.

The vestibule is the center for static (stationary) equilibrium. Hair cells similar to those in the organ of Corti are embedded in a gel-like material. Gravity weights the material, which pulls the hair cells downward. Since the vestibule always knows which way is down, it provides the sense of **static equilibrium**, or balance when standing still.

The semicircular canals also have hair cells in tubes filled with fluid. Each time we move, the fluid moves around the curved canal and bends the hair cells. The semicircular canal that is most parallel to our direction of movement will have its hair cells bent the most by moving fluid. This contributes to our sense of **dynamic equilibrium**, or balance while in motion.

Why do you think we have three semicircular canals?

____

____

____

ACTIVITY: Studying the Structure and Function of the Ear

Find the following structures on Figures 10.4 and 10.5 and the human ear models.

- Pinna
- Tympanic membrane (eardrum)
- Malleus
- Stapes
- Incus
- Vestibule
- Vestibulocochlear nerve
- Eustachian tube
- Cochlea
- Semicircular canals

Hearing Tests

Localization

Your lab instructor will demonstrate the proper technique for striking a tuning fork. Work in pairs.

1. Have your lab partner close his or her eyes and place a vibrating tuning fork six inches from the following positions: directly above the head, in front of the nose, behind the neck, right ear, left ear.
2. Repeat several times in random order. Where was your partner least able to localize the position of the sound source? ______________________. Propose a reason for this.

____

____

____

Bone Conduction

1. Strike the tuning fork and place the *base* on your mastoid process (the bony bump behind your ear). You should be able to hear the sound of the tuning fork as it vibrates the skull around your inner ear.
2. When the sound disappears, move the *tongs* of the tuning fork near your pinna. Do you hear the sound again? ______________________.
3. Reverse the process by striking the tuning fork and placing the tongs near your pinna. When the sound disappears, place the base of the tuning fork on your mastoid process. Do you hear the sound again? ____________. If you answered yes to this question, you may have a hearing impairment related to your conduction pathway that you bypassed when placing the base of the tuning fork on your mastoid process.

Frequency

Work in pairs. Have your partner close his or her eyes and place a vibrating tuning fork near either ear. Strike a tuning fork of a different frequency near the same ear. Did your partner correctly identify the second tuning fork as making a sound with a higher or lower pitch? __________. Did both sounds stimulate the same hair cells in the cochlea? Explain.

____

____

____

Balance Test

Have your partner time how many seconds you can stand on one foot with your eyes open (up to two minutes) versus with your eyes closed.

Open: ____________. Closed: ______________.

Considering that balancing requires a great deal of motor coordination (recall the role of the cerebellum), what does this experiment prove about the sense of balance?

____

____

____ ■

Study Questions

1. List four different kinds of receptors and state their basic function.

____

____

____

____

2. How is a receptor different from other neurons studied thus far?

____

____

____

3. What structures must light pass through before reaching the retina?

____

____

____

____

4. What do the terms "20/60 vision" and "20/15 vision" mean?

____

____

____

____

5. What is the transparent portion of the sclera called?

____

6. What are the two muscular modifications of the choroid, and what are their jobs?

____

____

____

____

7. Name four items that protect the eye from damage.

____

____

____

____

8. Which light-bending structure can change shape?

____

What happens when this structure loses its elasticity with age?

____

____

____

9. State the basic functions of the inner ear and list the structures involved in each function.

____

____

____

____

10. Briefly explain the difference between static equilibrium and dynamic equilibrium.

____

____

____

____

11. What structures are common to the hearing and equilibrium functions of the inner ear?

____

____

____

EXERCISE

The Endocrine System

OBJECTIVES

After completing this exercise, you should be able to:

1. Know the main glands of the endocrine system and their hormones.
2. Understand the general characteristics of major endocrine glands and hormones.
3. Be able to identify the major endocrine glands on models, specimens, and microscope slides.
4. Understand the basic function of the hormones discussed in this chapter.

MATERIALS

- ❑ Fetal pigs
- ❑ Dissecting equipment
- ❑ Human torso models
- ❑ Human brain models
- ❑ Reference charts
- ❑ Prepared microscope slides of:
  - ❑ Pituitary
  - ❑ Pancreas
  - ❑ Ovary
  - ❑ Testis
  - ❑ Thyroid

Introduction

Closely related to the nervous system in both function and chemistry, the endocrine system consists of glandular tissues that release **hormones**, chemical messengers carried by the blood to communicate with numerous body **target cells**. Target cells have **receptors** for these hormones so that they can interpret the message delivered by the hormone. Because endocrine glands are not limited by the need for a duct to carry their secretions, they can affect numerous, widely dispersed cells, tissues and organs, but work somewhat slower than nervous tissues.

Hormones

Peptide hormones are built from amino acids and closely related to neurotransmitters. They generally plug into receptors on the surface of target cells. **Steroid hormones** are based on a version of the cholesterol molecule (a lipid). These "fat-soluble" hormones easily cross the plasma membrane and pass on their message almost directly to the DNA of target cells via receptors in the nucleus.

This chapter covers the main endocrine glands and their hormones, but bear in mind that other hormones and endocrine structures exist. Tissues found in the digestive tract secrete several hormones that regulate digestive processes. The kidneys (and even the heart) release hormones that affect blood pressure, blood cell production, and fluid and electrolyte balance. Nearly every tissue in the body can release local hormones that trigger pain receptors, and can stimulate homeostatic responses in nearby tissues.

Endocrine System Control

For an endocrine gland to do its job, it needs to know exactly when to release its hormones. There are three ways that an endocrine gland may be triggered to release a hormone.

1) The gland's cells may be sensitive to levels of the chemical that it is supposed to regulate, as when high blood glucose levels trigger insulin release. This mechanism is called the **humoral control mechanism**.

2) The gland may be stimulated by the nervous system, as occurs when the sympathetic division of the ANS triggers epinephrine release.

3) The gland may be controlled by hormones of other glands, as when pituitary hormones stimulate the ovaries.

Major Endocrine Glands

Although the endocrine system includes many hormone-producing structures, we will limit this exercise to a discussion of the major endocrine glands of the human body (Figure 11.1). The major endocrine glands and the hormones they produce are listed in Table 11.1 on pages 98 and 99.

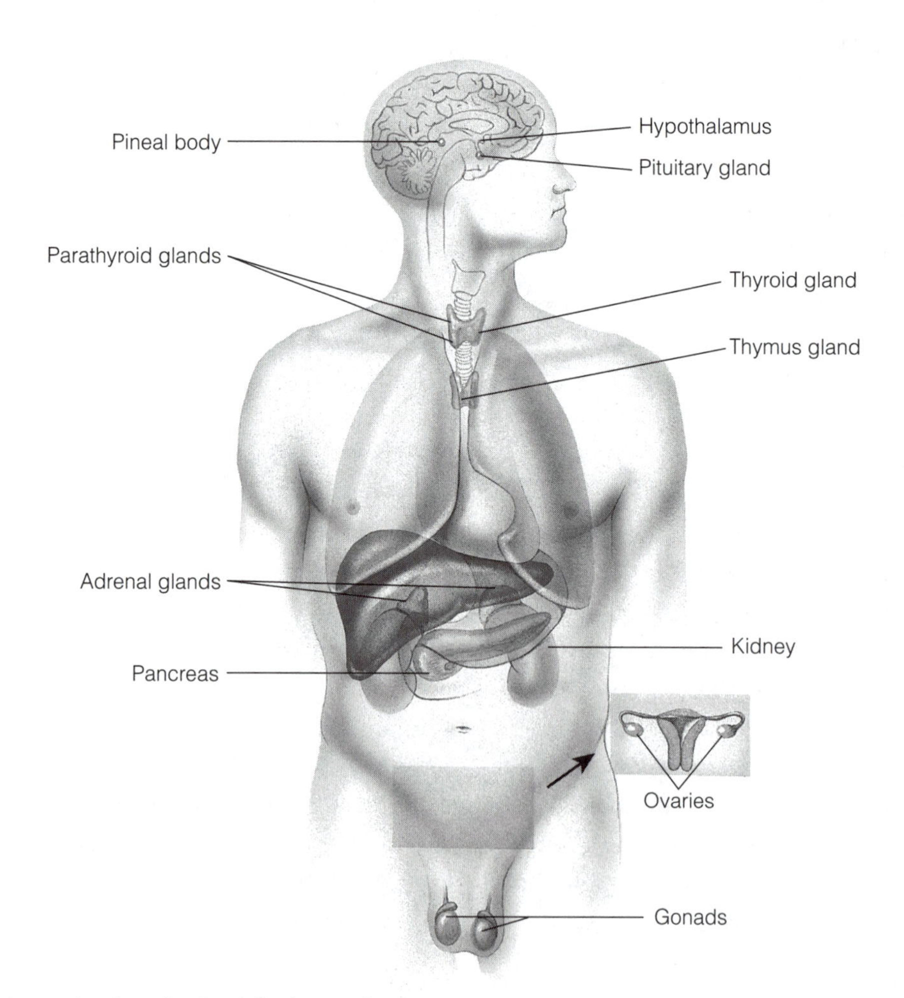

Figure 11.1 Major endocrine glands of the human body.

The Hypothalamus

There is a special relationship between the hypothalamus, the pituitary gland, and the rest of the endocrine system. Despite the pituitary's reputation as the "master gland," the hypothalamus is really in charge, tightly controlling pituitary function.

Control of the Posterior Pituitary

The hypothalamus directly controls the release of hormones stored in the posterior pituitary. The posterior pituitary is actually modified nervous tissue that is the storage and release site for hormones made in the hypothalamus.

Control of the Anterior Pituitary

The hypothalamus is indirectly connected to the anterior pituitary by the **hypophyseal portal system**. This system uses a few short veins to connect capillaries in the hypothalamus with a second set of capillaries in the anterior pituitary. Hypothalamic hormones produced in small amounts travel quickly down to the anterior pituitary to control hormone release. These hypothalamic hormones are either stimulating or inhibiting hormones, in that they either stimulate or inhibit the release of anterior pituitary hormones (see Table 11.1 for examples).

Posterior Pituitary

The posterior pituitary is unlike most other glands in that it is made of nervous tissue and is an outgrowth of the hypothalamus. It is the place where neurons from the hypothalamus end, and is the release site for some hormones made in the hypothalamus. The posterior pituitary releases two important hormones. **Oxytocin** causes contraction of the smooth muscle of the uterus and contractile cells of the mammary glands (milk letdown). **Antidiuretic hormone (ADH)** facilitates water reabsorption of the kidney tubules and causes smooth muscle contraction of arterioles in the systemic circulation, thus increasing blood pressure.

Anterior Pituitary

The anterior pituitary is true glandular tissue that makes its own hormones, but release of these hormones is still controlled by hypothalamic hormones. Hormones made in small quantities in the hypothalamus are carried directly to the anterior pituitary by the **hypophyseal portal system**. There are several important hormones produced by the anterior pituitary.

Growth hormone (GH) encourages growth of most body cells. Muscle, nerve, and cartilage- and bone-forming tissues (such as the epiphyseal plate) are particularly sensitive to GH. GH also helps cells to use available energy sources (carbohydrates and fats) and encourages protein synthesis. **Prolactin (Prl)** prepares the mammary glands for active milk production and secretion.

GH and Prl affect specific target tissues in the body. The remaining anterior pituitary hormones are **tropic** hormones, which affect other endocrine glands. Tropic hormones are typically named for their function. **Adrenocorticotropic hormone (ACTH)** stimulates the adrenal cortex (not the medulla) to release mainly glucocorticoids. **Follicle-stimulating hormone (FSH)** promotes oocyte development and stimulates growth of follicle cells in the ovary. In males, FSH promotes sperm production. **Luteinizing hormone (LH)** stimulates the follicle cells to produce estrogen, and plays a role in inducing ovulation. LH also converts the exploded follicle into the corpus luteum, a powerful tissue that produces progesterone and estrogen. In males, LH is called ICSH and stimulates testosterone release. **Thyroid-stimulating hormone (TSH)** stimulates certain thyroid cells to produce and release thyroxine.

Thyroid Gland

The thyroid gland is located just inferior to the larynx. **Follicular** cells are stimulated by TSH from the pituitary, and produce **thyroid hormone (TH)**. The raw materials needed to make TH, such as iodine and proteins, are located in the open spaces or **follicles**. TH promotes normal metabolic activity in most cells in the body. A second hormone of the thyroid, **calcitonin**, stimulates bone cells to transfer calcium from the blood into bone tissue. High blood Ca^{++} levels trigger release of calcitonin.

Table 11.1 Major Endocrine Glands and Their Hormones

Gland	Hormone(s)	Functions
Hypothalamus (Partial list of hormones)	Growth hormone–releasing hormone and growth hormone–inhibiting hormone (GH-RH) Gonadotropin-releasing hormone and gonadotropin-inhibiting hormone (Gn-RH) Thyrotropin-releasing hormone (TRH)	Now understood to be master of the "master gland" (the pituitary), hypothalamic hormones travel the short distance through the hypophyseal portal system to tightly control release of the anterior pituitary's hormones. These hormones are named for their functions.
Anterior pituitary	Growth hormone (GH)	Increases metabolism and growth of cells, especially bone and muscle.
	Prolactin	Stimulates milk-producing glands during pregnancy.
	Thyroid-stimulating hormone (TSH)	Stimulates the thyroid gland to produce and release thyroxine (thyroid hormone).
	Follicle-stimulating hormone (FSH)	Stimulates follicles in the ovary to grow and produce estrogen. Also promotes male sperm production.
	Luteinizing hormone (LH)	Stimulates ovulation and repair of the ruptured follicle into a corpus luteum, which produces progesterone and estrogen. Also promotes male testosterone production.
	Adrenocorticotropic hormone (ACTH)	Stimulates the adrenal cortex to produce and release hormones.
Posterior pituitary	Oxytocin	Stimulates smooth muscle contraction in the uterus during childbirth and in the mammary glands for milk letdown.
	Antidiuretic hormone (ADH)	Stimulates certain kidney tubules to reabsorb water; promotes water conservation and maintains BP.
Thyroid	Thyroxine or thyroid hormone (TH)	Increases metabolic rate in most cells; promotes growth and increases body temperature.
	Calcitonin	Stimulates bone cells to deposit calcium in bone; thus lowering blood calcium levels.

Parathyroid Glands

The parathyroid glands are a collection of 4 or, more often, 6 small glands that appear as rounded pieces of tissue protruding from or embedded in the posterior portion of the thyroid. **Parathyroid hormone (PTH)**, the antagonist of calcitonin, encourages bone cells to break up bone matrix in order to liberate calcium. PTH also tends to increase production of vitamin D and stimulates absorption and reabsorption of Ca^{++} from the intestines and kidneys.

Adrenal Glands

As with the pituitary, the adrenal glands are an embryological hybrid, derived from nervous tissue (medulla) and epithelial tissue (cortex).

Adrenal Cortex

The outer portion or cortex of the adrenal gland secretes steroid-based hormones in three distinct tissue layers.

1) The outer **zona glomerulosa** (rounded clumps of cells) secretes **aldosterone,** the main mineralo-

Table 11.1 Major Endocrine Glands and Their Hormones *(continued)*

Gland	Hormone(s)	Functions
Parathyroids	Parathyroid hormone	Stimulates bone cells to break down the bone matrix and liberate calcium from bone; increases blood calcium levels.
Adrenal medulla	Epinephrine/norepinephrine	Released during sympathetic stimulation, these hormones enhance the sympathetic response (i.e., increase heart rate and BP).
Adrenal cortex	Glucocorticoids (i.e., cortisol)	Stimulate cells to produce glucose from other organic molecules (mainly proteins); reduce inflammatory response.
	Mineralocorticoids (i.e., aldosterone)	Stimulate certain kidney tubules to reabsorb sodium and secrete potassium; increase water retention and BP.
	Androgens	Too small to have much effect in males; responsible for some female secondary sex characteristics.
Thymus	Thymus hormone (thymosin)	Stimulates maturation of the immune system's T-cells.
Pancreas	Insulin	Stimulates cells (mainly muscle, adipose, and liver) to absorb and store glucose as glycogen or convert it to fat, thus lowering blood sugar levels.
	Glucagon	Stimulates liver cells to break down glycogen and release glucose, thus increasing blood sugar levels.
Testis	Testosterone	Responsible for the growth and maintenance of male reproductive structures and secondary sex characteristics; may play a role in pituitary feedback loops involving sperm production.
Ovary	Estrogen	Hormone most responsible for female secondary sex characteristics.
	Progesterone	Partially responsible for some female secondary sex characteristics; promotes conditions for pregnancy. With estrogen, progesterone helps regulate the female cycle.

corticoid (mineral controlling adrenal hormone). Aldosterone stimulates kidney tubules to reabsorb sodium and excrete potassium. Although we tend to think of aldosterone as a hormone released to retain sodium and increase blood pressure, a high potassium level is actually the strongest trigger of aldosterone release.

2) The middle **zona fasciculata** (long strands of cells) secretes **glucocorticoids**, including cortisone, cortisol, and corticosterone. The glucocorticoids increase glucose production (for example, from amino acids) and stimulate adipose cells to break down lipids for energy. They also have anti-inflammatory properties and suppress the immune system. The main trigger seems to be ACTH from the adenohypophysis, which is released when the brain perceives danger, stress, or other adverse conditions.

3) The cortex's innermost area, the **zona reticularis** (irregular strands of cells) contains cells that secrete **androgens**. The small amount of androgens produced here are insignificant in males, but are significant in stimulating some of the secondary sex characteristics in females.

Adrenal Medulla

The inner adrenal medulla is made from nervous tissue. Directly connected to the nervous system by sympathetic nerves, it is rapidly stimulated during the fight or flight response. The medulla secretes **epinephrine** and **norepinephrine**. These hormones, commonly known as adrenaline, dilate the coronary arteries and increase blood pressure by stimulating vasoconstriction. They also increase heart rate and dilate the bronchial tree. Many drugs such as nasal decongestants and bronchial dilators are "adrenaline" mimics.

Pancreas

The pancreas is mostly an enzyme-producing gland of the digestive system, but is also dotted with clumps of hormone-producing cells called the **islets of Langerhans**. These cells produce **insulin**, which allows other cells to absorb glucose from the blood. Insulin especially stimulates liver, muscle, and adipose cells to absorb large amounts of glucose for processing. The liver and muscles store the glucose as the starch **glycogen,** while adipose cells store the energy from glucose as fat. Insulin is released when blood glucose levels are high. **Glucagon**, the antagonist of insulin is also secreted by other cells in the Islets of Langehans. It stimulates the liver to break up glycogen and release glucose into the blood.

Gonads

The gonads produce steroid hormones. The **testes** are mainly sperm-producing glands with numerous **seminiferous tubules**. The **interstitial cells** (also called cells of Leydig) lie in small clumps between the seminiferous tubules and produce **testosterone**. Testosterone is responsible for stimulating the growth of the male reproductive structures. During puberty, testosterone promotes the development of the male secondary sex characteristics, including facial hair and, with GH, additional muscle growth. The **estrogens** secreted by the **ovaries** are responsible for the female secondary sex characteristics, such as mammary glands and the tendency to deposit subcutaneous fat. **Progesterone** is also secreted during a portion of the menstrual cycle, and plays a major role in preparing the uterus for pregnancy.

Other Hormone-Producing Structures

The **thymus** secretes **thymosin**, which plays a major role in the maturation of T-lymphocytes, a critical portion of the immune system. The **pineal body** or gland secretes **melatonin**, but more importantly, may be part of the biological clock that initiates puberty. Although not often thought of as endocrine glands, the **kidneys** produce **renin,** which converts angiotensinogen to angiotensin (a hormone that plays a role in increasing blood pressure). The kidneys also produce **erythropoietin**, which stimulates erythrocyte production.

ACTIVITY: Dissecting the Fetal Pig

Place the fetal pig on a dissecting tray ventral (belly) side up. Use string to tie the right hind leg around the ankle. Run the string around the underside of the tray, and tie the left hind leg. Repeat for the forelegs. Carefully follow the steps listed.

1. Using a sharp scalpel, make a small incision through the abdominal skin *and* muscle about 1/2 inch above the umbilical cord. *Important:* Do not use the scalpel for further dissection work today. Use *scissors* to continue cutting along the midsagittal line on the ventral surface, first cutting upward toward the neck (it will be necessary to cut through the ribs as well).

2. Turn the tray around and cut down to the caudal (tail) end of the pelvic region, *leaving a 1/2-inch border around the umbilical cord.* This step is important to prevent cutting of the umbilical vein and arteries in the abdominal cavity. If you have a *female* pig, simply continue the midsagittal cut down into the pelvic region. Cut around the other side of the umbilical cord, again leaving about a 1/2-inch border. Stop your cut about one inch short of the anus. See Figure 11.2.

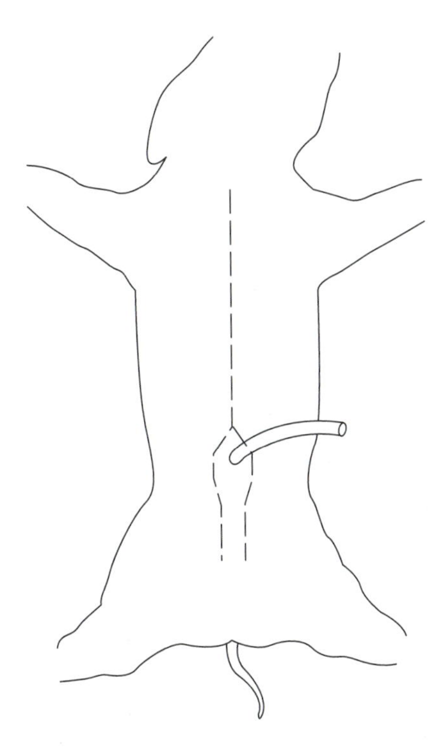

Figure 11.2 Midsagittal cut of the fetal pig dissection.

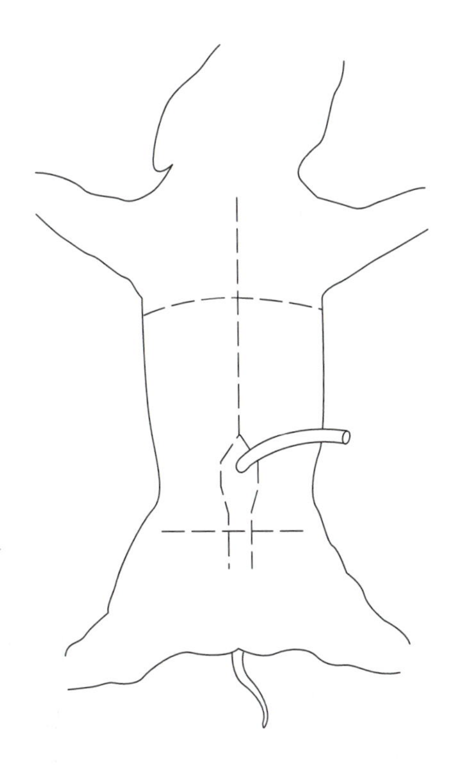

Figure 11.3 Transverse cuts of the fetal pig dissection (male cuts shown).

3. If you have a male pig, continue your cut down into the pelvic region slightly off-center. Cutting off-center ensures you do not cut the penis, which is incompletely formed in the fetal pig and appears as a thickened tube within the skin of the lower abdominopelvic area. Cut around the other side of the umbilical cord, again leaving about a 1/2-inch border, and then continue the cut off-center down into the pelvic area. Stop your cut about one inch short of the anus. See Figure 11.3.

4. Make two transverse cuts just below the forelegs and just above the hind legs. See Figure 11.3.

5. Lifting the flaps of skin and muscle on each side, cut the diaphragm, which is attached to the inside body wall. You should now be able to peel open the left and right flaps of the ventral body wall like a book.

6. If your pig has a great deal of brownish material (mainly bile and clotted blood), you may rinse the inside of the pig before attempting further investigation.

7. You may loop the loose string around the legs as necessary to get a better view into your specimen. When you are finished with your pig, *do not* cut the string! Simply slide the string from under the pan so that next time you need not retie the pig.

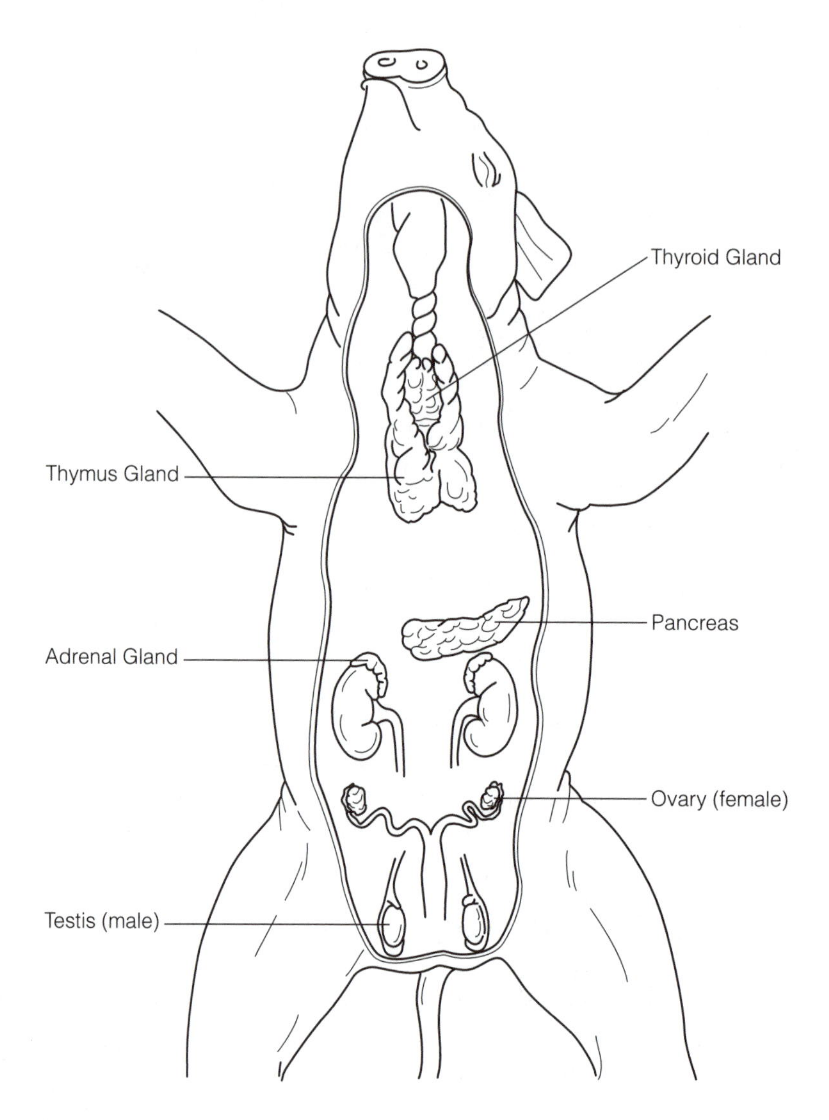

Figure 11.4 Major endocrine glands of the fetal pig.

Finding the Major Endocrine Glands in the Fetal Pig

Find in the fetal pig the major endocrine glands listed below. Use Figure 11.4 for reference.

- **Thymus** The thymus is the light-tan-colored gland that partially covers the heart and extends upward along the lateral portion of the trachea. It is the easiest gland to find, as it is usually visible without further dissection.
- **Thyroid** Usually brownish in color, the thyroid is a small round or oval-shaped gland just below the larynx in the neck. The thyroid is usually covered by a thin layer of glandular tissue that must be carefully pulled away.
- **Pancreas** Lift up the stomach. Underneath it you should see a thin, semi-transparent membrane covering the loose glandular tissue of the pancreas. Carefully break and pull the membrane. You should see the bumpy, tan-colored pancreas underneath, looking a bit like a miniature ear of corn.

- **Adrenal glands** These glands can be difficult to find. In the human, the paired adrenal glands lie on top of the kidneys. In the pig, they are superior and medial to the kidneys. It is probably best to look for the right adrenal gland (the *pig's* right). Push the intestines and all loose organs over to the *pig's* left side (do not cut them out!). Find the brownish kidney partly covered by the same semi-transparent membrane that covers the pancreas. Gently pull the membrane from the superior/medial part of the kidney. You should see a lighter-colored "cap" that may now be partially disconnected from the kidney. This is the adrenal gland.
- **Testes/Ovaries** The testes have not descended in the fetal pig and can be found by lightly slicing the skin and muscle covering the lower pelvic cavity about one inch laterally from center. Push the external area that looks like the developing scrotum of the pig. You should see a bean-sized mass of soft tissue move up and down as you palpate. Use your forceps to uncover the small, brownish, oval testis in this mass of soft tissue. The ovaries float near the border of the lower abdominal and upper pelvic cavities. Simply lift the intestines on either side and look for the small, round, brownish ovary.

Endocrine Glands in the Human Models

Identify the pituitary gland on the sheep brain and human brain models. Where possible, find the major endocrine glands on the other models that may be provided by your instructor. ■

Histology

As directed by your instructor, view the slides either on your own or as set up on demonstration.

- **Pituitary gland** View a slide of the pituitary gland under low power. Note the two very different tissue types. The mass with the "streaked" texture is the posterior pituitary. The streaked texture is caused by these axons in the tissue. The "dotted" mass is the anterior pituitary. The dots are the nuclei of cuboidal epithelial cells. Use higher power to view these axons and cells.
- **Pancreas** The pancreas is mostly an *exocrine* gland, so most of the cells you see are acinar cells involved in digestive enzyme production. The *endocrine* cells are clustered in clumps called islets of Langerhans. Under low power, scan for somewhat rounded clumps of cells that may stain a little lighter than most pancreatic cells. Also view under higher power.
- **Ovaries** The follicular cells are easy to find since they surround the large, round oocytes in the ovary. A mature follicle will be a hollow spherical mass of cells with or without an egg cell on a stalk in the middle. A corpus luteum is a large, light-staining mass of cells found in the ovary during the second half of the female cycle and during pregnancy.
- **Testis** The testis is mainly an exocrine gland, responsible for sperm production. But the interstitial cells, or cells of Leydig, produce testosterone. In cross section, the interstitial cells appear as small clumps of cells wedged in between the large, round seminiferous tubules.
- **Thyroid** Under low power, students often mistake this slide for poorly stained adipose tissue. Those big, light-pink blobs are not cells, but thyroid follicles that contain the raw materials for making thyroid hormone. Under higher power, you will see that the follicles are surrounded by a single layer of small follicular cells, which produce thyroid hormone. The other cells that are not around a follicle produce the hormone calcitonin.

Study Questions

1. What are the two chemical classes of hormones?

    ____

    ____

2. The hormones released by the posterior pituitary are actually made in the ______________.

3. What are tropic hormones, and where are they made?

____

____

____

____

4. Fill in the name for the endocrine portion of each of the following endocrine glands:

Testis _______________.

Pancreas _______________.

Ovary _______________.

5. Two endocrine glands discussed in this chapter have both a portion made of glandular (epithelial) tissue, and a portion made of modified nervous tissue. What are they?

____

____

Which portions of each gland are made of epithelial tissue?

____

____

Nervous tissue?

____

____

6. Match the endocrine gland with its hormone:

Pancreas ____	a. growth hormone
Anterior pituitary ____	b. epinephrine
Ovary ____	c. insulin
Adrenal medulla ____	d. estrogen

7. Match the hormone with its function:

Glucocorticoids ____	a. Stimulates liver cells to break down glycogen and release glucose.
Glucagon ____	b. Stimulates smooth muscle contraction in the uterus during childbirth.
ACTH ____	c. Stimulates the adrenal cortex to produce and release hormones.
Oxytocin ____	d. Stimulate cells to produce glucose from other organic molecules.

8. In terms of both identification and function, what is the difference between an ovarian follicle and a corpus luteum?

____

____

____

____

9. Predict what would happen to a person who was unable to produce ADH.

____

____

____

____

10. Predict what would happen to a person who was unable to produce sufficient thyroid hormone.

____

____

____

____

EXERCISE

The Blood

OBJECTIVES

After completing this exercise, you should be able to:

1. Name the two major components of whole blood.
2. Understand the functional importance of plasma.
3. Define "formed elements" and describe the major functions of each element.
4. Identify each of the formed elements using a prepared microscope slide, appropriate illustrations, or both.
5. Discuss and understand the basis for blood types.

MATERIALS

- ❑ Sheep or ox blood
- ❑ Capillary tubes
- ❑ Clay
- ❑ Hematocrit kits or scales
- ❑ Centrifuge
- ❑ Compound light microscope
- ❑ Blood smear with Wright's stain
- ❑ Medical dictionary
- ❑ 4 blood-typing slides
- ❑ 8 toothpicks

Shared materials:

- ❑ 4 unknown simulated blood samples
- ❑ Anti-A simulated typing serum
- ❑ Anti-B simulated typing serum

Introduction

Your body is composed of many different types of cells, most of which are highly specialized structures capable of performing specific functions. Specialization, however, carries with it some drawbacks. The cells of multicellular organisms are not capable of independent existence. These cells cannot search for food, nor can they run away from their own battles. Something else has to tend to all these functions. That something is, for the most part, blood. Your blood is a tissue that communicates with the interstitial (between cells) fluid, which bathes the rest of the body cells via capillaries. Capillaries are thin-walled tubes (only one cell thick) that connect the arterial circulation to the venous circulation. Since the capillaries are so thin, fluid, and small, molecules freely move back and forth across the capillary wall. In this way cells communicate with the blood to exchange waste products and nutrients. Your blood is their link to the outside world.

Fluid Portion of Blood

The fluid portion of the blood is called **plasma.** It constitutes 55–60% of the blood by volume. Dissolved in plasma are several different kinds of proteins (albumin, globulins, and clotting proteins), as well as ions, nutrients, and hormones. When all the clotting proteins have been removed from plasma, it is called **serum**. Over 95% of the plasma is water, but the substances present in the fluid medium have some very important functions, such as:

- Maintaining blood volume
- Maintaining proper blood viscosity (thickness)
- Providing buffers to guard against pH changes
- Providing material to aid the clotting process
- Transporting nutrients, wastes, and hormones

Formed Elements of the Blood

Refer to Figure 12.1 when studying the formed elements of the blood.

Erythrocytes

When you think of blood, you think of the color red. The redness of blood is due to **hemoglobin**, a protein contained inside the **erythrocytes** (red corpuscles). *Heme* of the hemoglobin is the portion that contains iron, to which oxygen attaches. *Globin* is the protein. The function of the hemoglobin is to transport the gases oxygen and carbon dioxide. The shape of the erythrocyte (*erythro* = red; *cyte* = cell) is suited to its function. It is normally a bioconcave disc with no nucleus, thus allowing maximum packing volume and surface area for its transport function.

All of the cellular elements of blood develop in the red bone marrow. There are approximately 4.5–5.5 million erythrocytes per cubic millimeter (mm^3) of blood. Normally, cell production equals cell destruction. When one or the other is out of balance, abnormal function may result. **Anemia** is a general term that simply means a deficiency of oxygen-carrying capacity. Here are two common types of anemia:

- Iron-deficiency anemia occurs when a person does not ingest enough iron to keep up with the bone marrow's demands (remember the "heme" part of hemoglobin).
- Pernicious anemia occurs when a person cannot properly absorb enough vitamin B_{12} from the gastrointestinal tract. People with this condition can absorb vitamin B_{12} only when injected with it. Vitamin B_{12} is needed to manufacture erythrocytes.

ACTIVITY: Performing a Hematocrit

1. Obtain a capillary tube and transfer the blood sample provided into the tube until it is approximately 3/4 full. Stick the bottom of the tube into the clay to seal the tube at one end. Your instructor may wish to demonstrate the proper way to prepare a capillary tube.

2. Place the capillary tube in the centrifuge slot with the plugged end toward the outside. Close the centrifuge and spin on medium-low speed for three minutes. Note: Because there is great variety among centrifuges, your instructor should provide guidance on safe use of the centrifuge and the centrifugation speed appropriate for your equipment.

3. Turn off the centrifuge and when it stops spinning, carefully remove the capillary tube. Use the charts provided to determine the percentage of plasma vs. formed elements. Record the hematocrit erythrocyte/formed element percentage: _______. ■

(text continued on page 108)

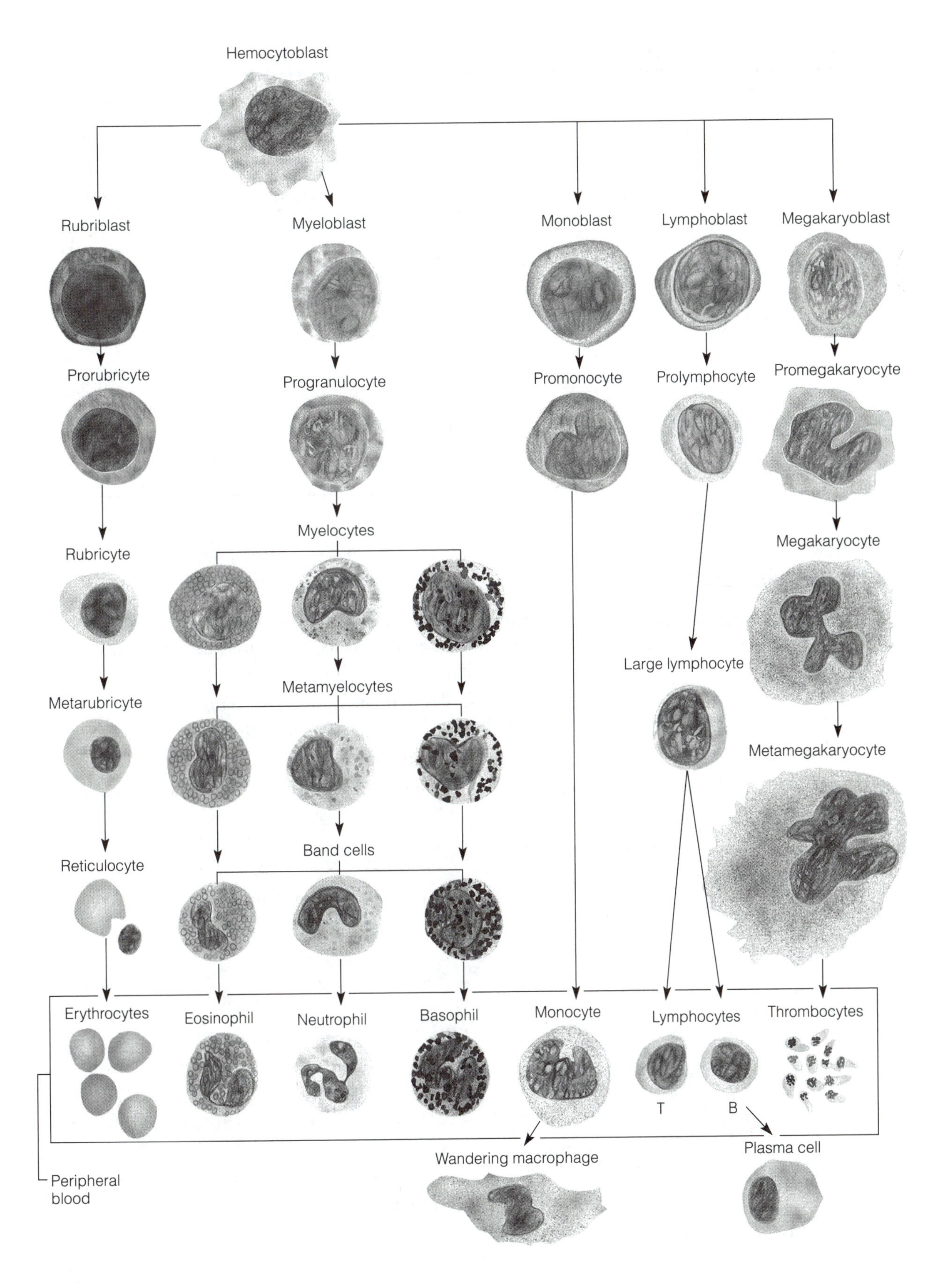

Figure 12.1 The morphology of human blood cells.

Table 12.1 Identifying Traits of Leukocytes

Leukocyte	Nucleus	Cytoplasm	Size
Neutrophils	two or more lobes connected by narrow strands	small granules that appear light pink to blue-black	12–14 µm
Eosinophils	bi-lobed nucleus, look like earphones	coarse granules that appear reddish-orange	0–14 µm
Basophils	S-shaped, hidden by large, darkly stained granules	very large, reddish-purple to blue-black granules	8–10 µm
Monocyte	oval, kidney bean–shaped or U-shaped	more cytoplasm than in a lymphocyte; gray-blue color	14–24 µm
Lymphocyte			
large	very large round nucleus	thin strip of clear blue cytoplasm	12 µm
small			8 µm

ACTIVITY: Observing Erythrocytes

1. Obtain a compound microscope and a prepared slide of smeared blood that has been stained with Wright's stain.
2. Focus one end of the smear under low power, then switch the magnification to high-dry power.
3. Examine the erythrocytes, keeping in mind that there are about a thousand times more red corpuscles than any of the other formed elements. Notice the size and shape of erythrocytes.

   Describe the appearance of the blood elements you are examining.

   Are the cells the same size or do their sizes differ?

   Keep your slide focused for the next activity: "Recognizing Leukocytes". ■

Leukocytes

All of the leukocytes (*leuko* = white; *cyte* = cell), or white blood cells, originate in the red bone marrow. Some travel outside the circulation and "seed" other areas of the body. There are normally 5,000 to 10,000 leukocytes per cubic millimeter of blood. These cells fall into two categories based on the presence or absence of obvious granules in the cytoplasm (Table 12.1).

Neutrophils

- The most common white blood cell (WBC)
- Main function is phagocytosis; your "first line of defense" against infection
- Very light, pale-pink granules when stained

Eosinophils

- Tend to increase in parasitic infections and allergies
- Can leave the bloodstream to colonize other areas of the body
- Have large red granules (when stained) in their cytoplasm

Basophils

- The least common WBC
- Have large blue (when stained) cytoplasmic granules which contain mostly histamine; therefore, basophils play a major role in allergic reactions and in inflammation

Lymphocytes

- Second most common WBC
- Involved in the immune response; highly specialized
- Function is manufacturing antibodies and eliminating anything foreign to the organism
- Are found in lymphatic organs (spleen, liver, lymph nodes)
- Certain lymphocytes are the "target" cells for the AIDS virus

Monocytes

- Very large, actively phagocytic cells
- Play a very large role in the immune response, but also are general scavengers, actively "gobbling up" small particles (like bacteria) that don't belong in the body. When monocytes leave the bloodstream, they are called macrophages (*macro* = large; *phage* = eating).

ACTIVITY: Recognizing Leukocytes

1. Continue examining the stained slide. This time, concentrate on the various larger cells with purple-stained nuclei.
2. Slowly move the slide in a regular pattern. Attempt to identify each type of leukocyte.
3. Use illustrations to help with your investigation. Eosinophils and basophils are relatively few in number. When you locate either of these cells, inform your instructor so your classmates have an opportunity to see them.
4. In the appropriate spaces below, draw each of the leukocytes as they appear to you.

Neutrophil

Eosinophil

Basophil

Lymphocyte

Monocyte

Hematologic determinations can provide some valuable clues about what is happening in an organism. One of the standard components of a physical examination is an examination of the quality of the blood. ■

Chart 12.1

Measurement	Normal Value	Clinical Significance of Increased Value	Clinical Significance of Decreased Value
Total Erythrocyte count (number per mm³)			
Total leukocyte count (number per mm³)			
Differential count			
Neutrophils			
Eosinophils			
Basophils			
Lymphocytes			
Monocytes			

ACTIVITY: Studying Common Hematologic Determinations

Use a medical dictionary or other reference text to complete Chart 12.1. ■

Platelets

Platelets are derived from large cells in the bone marrow that break into fragments. Each fragment is enclosed by a membrane, thus giving rise to a platelet. Inside the platelet are important proteins and other chemicals needed for the clotting process. Naturally, you don't want your blood to clot inside your blood vessels (this is called a thrombus), so normally the platelets are very stable as they circulate. When they encounter a "rough surface" (for example, tissue damage), they form a platelet plug and the clotting mechanism is initiated. Many of the chemicals that participate in the clotting process circulate freely in the bloodstream. They won't become active until the damaged platelets release the chemical thromboplastin.

In addition to calcium, plasma, and platelet proteins, vitamin K is also needed for normal blood clotting. Vitamin K helps the liver synthesize prothrombin.

The following is a simplified version of the clotting process:

Prothrombin (a plasma protein) —— Thromboplastin from platelets ——→ Thrombin (enzyme)

Fibrinogen (a plasma protein) —— Thrombin ——→ Fibrin (clotting protein)

Blood Typing

Every individual has unique "marker" chemicals that protrude from the cells. These marker chemicals are called "self antigens" and they label your body cells in such a way that your immune system recognizes these cells as belonging to you. Your red blood cells (RBCs) have many marker antigens on the surface of the cell membrane. In blood typing, you will be learning about two of them.

One type of marker antigen is a complex sugar molecule that can exist in several forms. Depending on which sugar molecule(s) you have on the cell surface, you are classified into Group A, Group B, Group AB, or Group O.

A second kind of antigen is a protein molecule called the **Rh antigen.** People who have this antigen are classified as Rh positive; those without it are Rh negative.

These inherited antigens exist on your RBC surface at the time of birth. Very shortly after birth, your body begins to manufacture antibody molecules directed against the other major blood group types of antigens. We do not generally manufacture antibodies against our own cells and their marker molecules, so the ABO antibodies we have would be typically opposite from our blood type. People classified as Group A will manufacture the anti-B antibody. Individuals classified as Group B will manufacture the anti-A antibody. Those with the blood type AB have both A and B chemicals projecting from the surface. These people will not manufacture any antibodies directed against the major blood groups. Those without A or B surface antigens are blood type O, and have both anti-A and anti-B antibodies.

Similarly, if you have the Rh antigen projecting from the RBC surface, you will not manufacture any antibody against the Rh antigen. However, if you do not have the Rh antigen, you will not naturally start manufacturing antibodies against the Rh antigen unless you are exposed in some way to Rh-positive cells.

Process of Blood Typing

Blood typing is performed with anti-sera containing high concentrations of anti-A and anti-B antibodies (agglutinins). Blood samples are mixed with anti-sera containing known antibodies. When the antibody reacts with the like antigen (agglutinogen) on the RBC, an agglutination (clumping) occurs.

For example, if agglutination occurs with the anti-A serum, the blood type is A. If agglutination occurs only with the anti-B serum, the blood type is B. If both anti-sera produce clumping, the blood type is AB. The absence of agglutination indicates blood type O.

ACTIVITY: Determining ABO Blood Type

Each team of two students will determine the blood type of each of the four unknown blood samples.

1. Label each of your four typing slides:
   - Slide 1: Mr. Smith
   - Slide 2: Ms. Jones
   - Slide 3: Mr. Green
   - Slide 4: Ms. Brown
2. Place 3–4 drops of Mr. Smith's blood in well A of Slide 1 and 3–4 drops in well B of Slide 1.
3. Prepare Slides 2, 3, and 4 in the same manner using the blood of Ms. Jones, Mr. Green, and Ms. Brown, respectively.
4. Add 3–4 drops of the simulated anti-A serum in each of the four "A" wells.
5. Add 3–4 drops of the simulated anti-B serum in each of the four "B" wells.
6. Add 3–4 drops of simulated anti-D (Rh) serum in each of the four "Rh" wells.
7. Using separate toothpicks, stir each sample of serum and blood. A positive test is a strong agglutination. It may take a couple of minutes to become evident.
8. Record your results in Table 12.2.

Table 12.2 Agglutination Reactions

Anti-A Serum	Anti-B Serum	Anti-Rh Serum	Blood Type
Slide 1: Mr. Smith			
Slide 2: Ms. Jones			
Slide 3: Mr. Green			
Slide 4: Ms. Brown			

Identify the antigen(s) present on the red cells of Mr. Green's blood.

____

____

Identify the antibody molecules that would be present in Mr. Green's plasma.

____

______________________________ ■

Study Questions

1. Look up and define:

   a. Polycythemia

   ____

   ____

   b. Leukocytosis

   ____

   ____

   c. Leukopenia

   ____

   ____

   d. Leukemia

   ____

   ____

2. When an erythrocyte has a shape other than that of a biconcave disk, how might its function be affected?

   ____

   ____

   ____

   ____

3. Considering its function, why is the biconcave disk shape of the RBC superior to a simple sphere?

4. How might the lack of a nucleus be an advantage to the red blood corpuscle?

How might it be a disadvantage?

5. What is the difference between a total leukocyte count and a differential count?

6. Where are RBCs produced?

WBCs?

Platelets?

7. How is blood viscosity related to blood circulation?

8. What is the danger to an individual with a:

Low total white blood cell count?

High total white blood cell count?

Low number of platelets?

Low hemoglobin determination?

Low red blood corpuscle count?

High red blood corpuscle count?

9. Why is it dangerous to supply a blood type different than the recipient's blood type in a transfusion?

EXERCISE

13 The Heart and General Circulation

OBJECTIVES

After completing this exercise, you should be able to:

1. Identify by observation and dissection the major structural features of the mammalian heart.
2. Compare and contrast the anatomy of an artery, a vein, and a capillary.
3. Locate major arteries and veins on a chart or model.
4. Understand systemic and pulmonary circulations.
5. Listen to sounds produced by the cardiac cycle.
6. Determine the heart rate and measure the blood pressure using a sphygmomanometer.
7. Observe the effects of exercise on heart rate and blood pressure.

MATERIALS

- ❑ Fetal pigs
- ❑ Entire preserved sheep heart
- ❑ Coronally sectioned sheep heart
- ❑ Alcohol swabs
- ❑ Blunt metal probe
- ❑ Bicycle ergometer
- ❑ Large human heart model
- ❑ Red and blue pencils
- ❑ 6 small human heart models
- ❑ Stethoscope
- ❑ Alcohol swabs
- ❑ Sphygmomanometer

Introduction

The heart is a hollow muscular organ. The contraction of the cardiac muscle creates the pressure that moves blood through the chambers of the heart and through all the vessels of the body. The heart is, however, more than a pump; it secretes hormones that help to regulate the blood pressure and the delicate balance of the body's fluid and salt.

Blood moves away from the heart in arteries and returns to the heart in veins. The arteries and veins are connected to each other by capillaries.

The right side of the heart receives blood that has circulated through the system and is low in oxygen (deoxygenated). This blood is moved from the right side of the heart to the lungs, where it becomes oxygenated. From the lungs, blood is returned to the left side of the heart, which circulates blood to all parts of the body.

The Anatomy of an Artery, Vein, and Capillary

Blood vessels of the human body vary in size and diameter (Figure 13.1). **Arteries** are large, thick-walled vessels that transport blood away from the heart. Blood is distributed to all parts of the body, from arteries to **arterioles** (smaller arteries) to capillaries. **Capillaries** are networks of microscopic vessels one cell thick. It is here that materials may move from the circulating blood to the body cells, or from the body cells to the blood. From the capillaries, blood returns through the **venules** (smaller veins) and flows to larger **veins.** Veins are thin-walled and have valves that keep the blood moving toward the upper chambers of the heart. All blood vessels except capillaries have smooth muscle in their walls. This allows changes in blood vessel diameter, thereby regulating blood pressure and blood distribution.

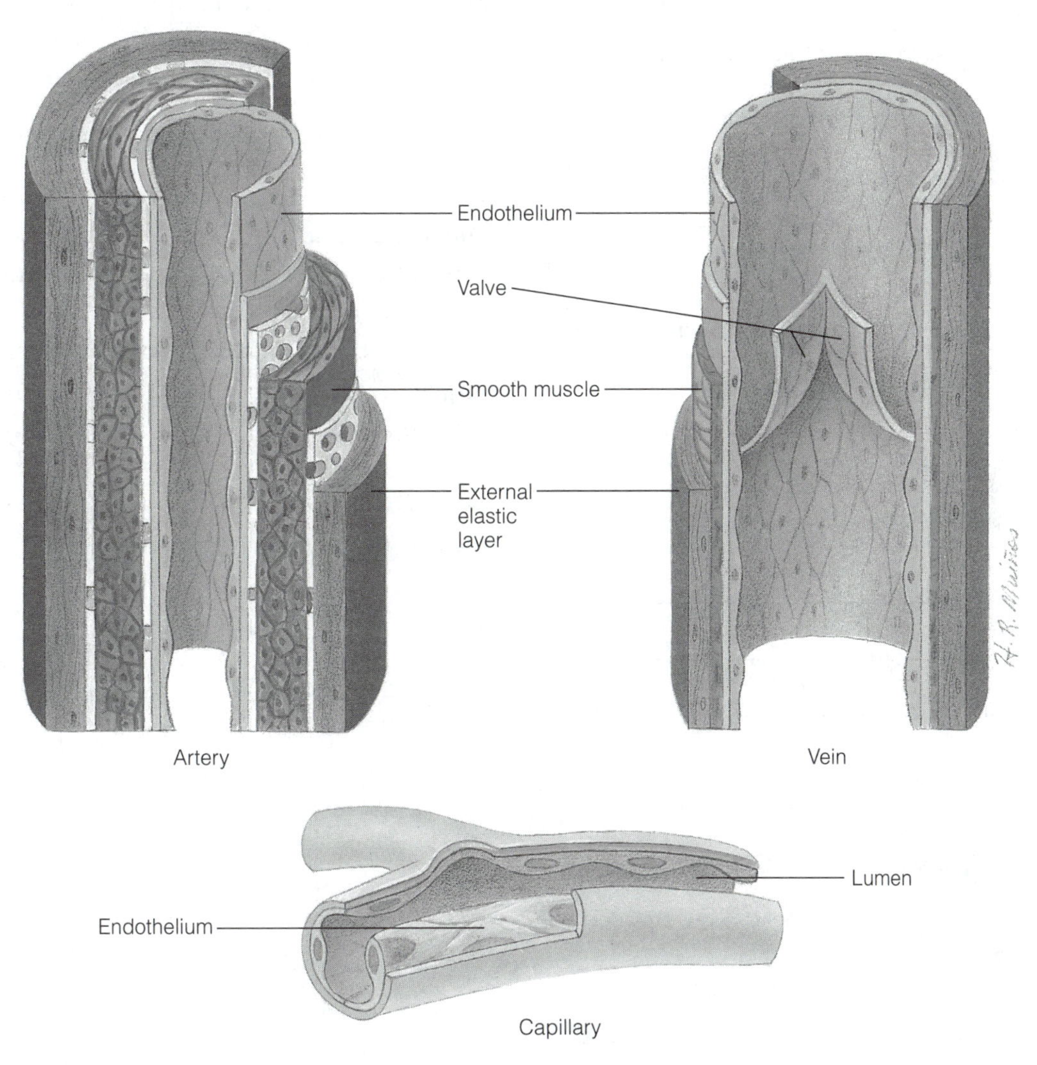

Figure 13.1 Comparison of blood vessels.

ACTIVITY: Studying Blood Vessels

Examine Figure 13.1 and Figure 13.2.

Capillaries form many branches that eventually make up a network. Capillaries are microscopic in diameter. Red corpuscles can be observed moving in single file through a capillary.

Physiologically, the rate of flow of blood through the capillary network is of great importance because these small vessels are responsible for the exchanges between blood and all cells of the body.

How do veins and arteries differ in structure and function?

____

____

____

____

____

____

How does the size and structure of a capillary relate to its function?

____

____

____

____

____

______________________________________ ■

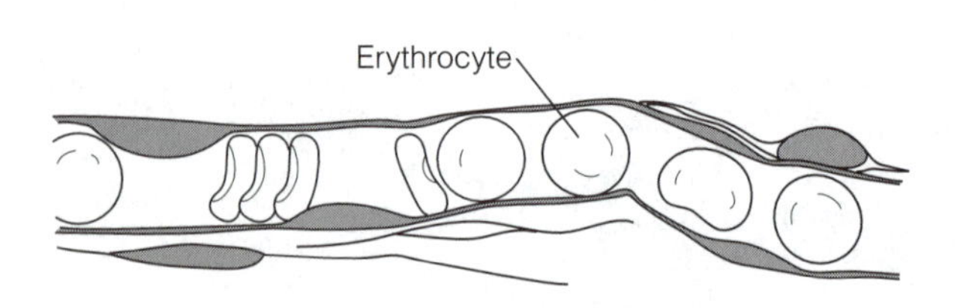

Figure 13.2 Capillary: longitudinal section with blood cells.

ACTIVITY: Structure of the Heart

External Anatomy

Using a plastic human heart model, sheep heart, and fetal pig, identify the structures in bold listed below. Look to Figures 13.3 and 13.4 for further reference.

The large vessels emerging from the heart are different in size and thickness. The two large arteries are the **aorta**, which transports blood away from the heart to all parts of the body, and the **pulmonary trunk**, which transports blood to the lungs. These vessels originate from the two lower chambers of the heart.

Two earlike flaps on either side of the heart called auricles extend from the upper chambers, the **right atrium** and the **left atrium.** They act as a safety device for the overflow of a sudden large blood return to the atria.

On the dorsal side are the two collapsed thin-walled veins, the **superior vena cava** and the **inferior vena cava.** These veins return blood low in oxygen to the right atrium. The four **pulmonary veins** return oxygenated blood to the left atrium.

The **pericardium,** a sac covering the heart, will be visible on the fetal pig heart only. Fat marks off and surrounds each of the four chambers of the heart. Embedded in the layer of fat are the coronary blood vessels, which carry the blood to and from the heart muscle, the **myocardium.**

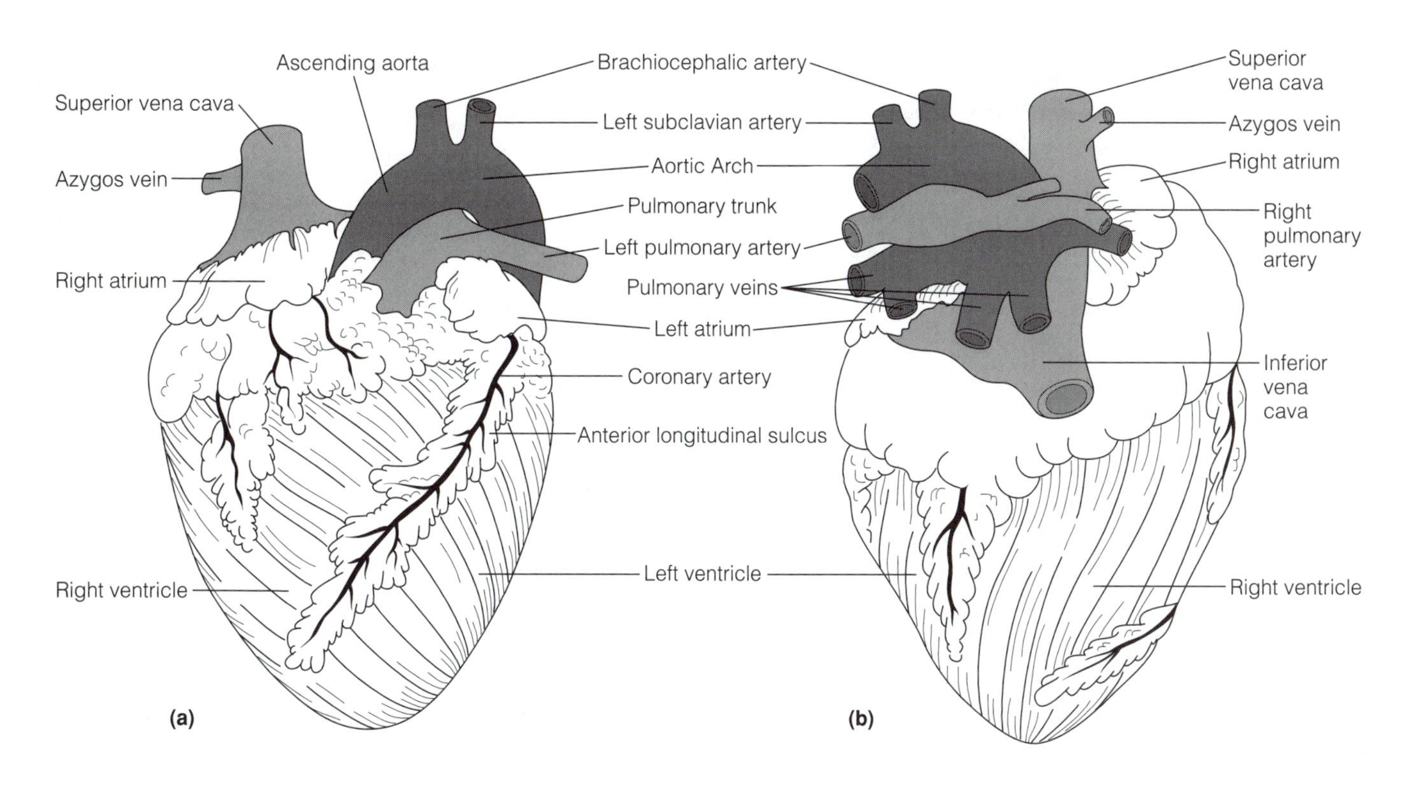

Figure 13.3 Sheep heart: a) ventral view b) dorsal view.

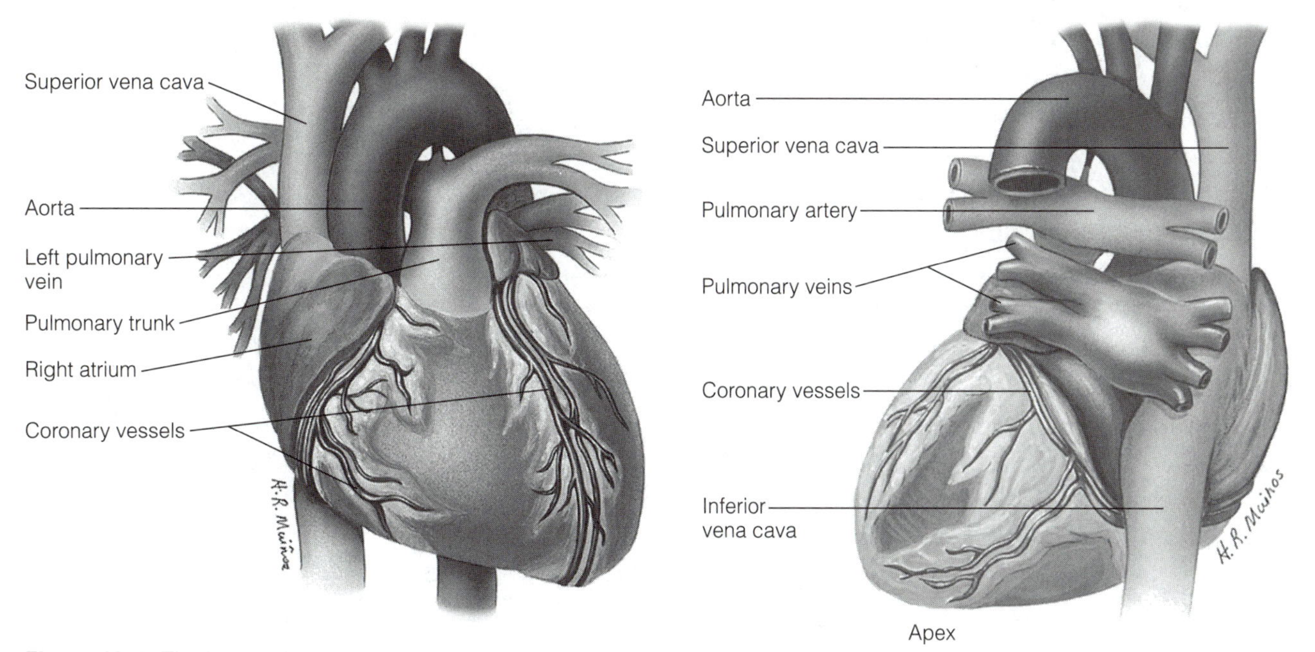

Figure 13.4 The human heart.

Internal Anatomy

Using the frontally sectioned sheep heart and human heart models, observe the internal features and structures defined below. Look to Figure 13.5 for further reference.

Open the sheep heart and human heart model and locate the chambers of the heart. Remember—the correct anatomical position will place the right-side structures on your left and the left-side structures on your right.

As you lift the anterior wall of the heart, you will see the two lower chambers, the right and left ventricles. Note the difference in the thickness of the right and left ventricular walls. The left ventricle generates sufficient pressure to pump the blood through the vessels of the body. The right ventricle pumps blood to the nearby lungs. A thick wall separates the two sides of the heart. In the adult, there is no connection between the right and left heart. The **right** and **left atria** are the smaller chambers above each ventricle.

Next, locate the thin tendinous cords attached to the base of the ventricles. These cords prevent the leaflike valves from turning inside out when the ventricles contract. The valves located between the atria and ventricles are referred to as **atrioventricular**. The **bicuspid** or **mitral valve** separates the left atrium and the left ventricle.

The **tricuspid valve** separates the right atrium from the right ventricle. These **atrioventricular** (AV) **valves** open as the atria fill with blood. They close to prevent any backflow of blood into the atria when the ventricles contract.

The valves located at the origin of the aorta and pulmonary artery are referred to as semilunar valves. Lift the anterior of the heart a little further to locate the **pulmonary semilunar valve.** You can find this valve where the pulmonary artery leaves the right ventricle. This valve has three semilunar cuplike sacs attached to the walls of the artery. When the right ventricle contracts, pressure is created against the sacs. The pulmonary semilunar valve opens by flattening against the vessel wall, and blood rushes into the pulmonary artery and moves on to the lungs.

The **aortic semilunar valves** and the **pulmonary semilunar valves** are found at the base of the aorta and pulmonary trunk respectively. You can locate the aortic semilunar valves at the exit of the aorta from the left ventricle. The opening and closing of these valves at proper intervals keeps blood moving in a circuit away from and then back to the heart. The pulmonary semilunar valves are not visible in the sheep heart without further dissection, but they are easily seen on the models and have the same function as the aortic semilunar valves.

Compare all of the features of the sheep heart with a human heart model. Refer back to Figure 13.3 and 13.4 for further reference.

- Right auricle
- Left auricle
- Left ventricle
- Right ventricle
- Superior vena cava
- Inferior vena cava
- Aorta
- Pulmonary artery
- Pulmonary veins

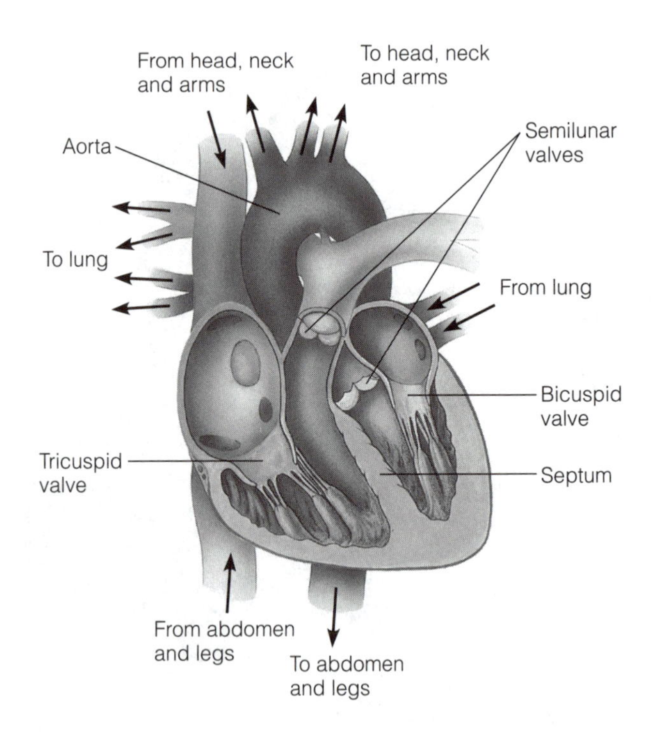

Figure 13.5 Human heart circulation.

Additional Blood Vessels

Also find the vessels previously described in the fetal pig and the human models provided. See also Figure 13.6. Using the pictures in your lab manual and other color plates provided in your laboratory classroom, find the following blood vessels:

- Brachiocephalic artery
- Left and right common carotid arteries
- Left and right subclavian arteries
- Thoracic and abdominal aorta
- Renal arteries
- External iliac arteries
- Jugular veins
- External iliac veins

If time permits, your instructor may also wish to point out structures unique to fetal circulation such as the umbilical arteries, umbilical vein, and ductus arteriosus.

Functional Anatomy

Contraction of the walls of a chamber is called **systole.** When the left ventricle contracts, blood is moved through the aorta and along to all the cells of the body. Relaxation of the walls of a chamber of the heart is called **diastole.** Blood that has circulated through the body is returned to the right atrium via the superior and inferior vena cava.

Pathway of Circulation

Learn the general circulatory path of blood into and out of the heart. Using Figure 13.5 as a guide, trace the path of blood, beginning with the superior and inferior vena cava and traveling along to the aorta. Color blue the areas of the heart and vessels of the heart transporting deoxygenated blood. Color red the areas of the heart and vessels containing oxygenated blood.

On Figure 13.4, pay special attention to the coronary vessels, which are so vital because they supply the myocardium with oxygenated blood. Without this supply of fresh blood, the heart muscle ceases to contract and the organ cannot fulfill its function.

Why is the heart called a double pump?

____

____

____

____

____

What are the main functions of the valves of the heart?

____

____

Which blood vessels supply the heart muscle with oxygenated blood?

____

______________________________ ■

Sounds of the Heart

As the heart beats, it makes "lubb-dupp" sounds that can be heard with a stethoscope. The "lubb" is caused by the snapping shut of the bicuspid and tricuspid valves. The "dubb" is the closing of the pulmonary and aortic semilunar valves. These sounds provide information about the condition of the valves. Heart murmurs are unusual sounds that often result from defects in the heart valves.

ACTIVITY: Observing the Sounds of the Heart

1. Clean the earpieces of the stethoscope with an alcohol swab and position the earpieces.
2. Place your finger in the notch at the base of the throat, and run your finger straight down to the angle of the breastbone. Immediately to the left is the second intercostal space.
3. Move your finger down until you reach the fifth intercostal space. This is where the heart lies closest to the body wall.

Chart 13.1

Method	Beats per Minute Partner	Yourself
Stethoscope		
Pulse		

4. Listen to the heart sounds by placing the bell of the stethoscope on the left side of the chest at the fifth intercostal space.
5. With a partner at rest, position the stethoscope and count the heartbeat for 30 seconds, then multiply by two.
6. Record your partner's heart rate in Chart 13.1.
7. Switch and have your partner determine your heart rate in a similar manner. Record your heart rate per minute in Chart 13.1. ■

Heart Rate

As a result of the heart beating, a rhythmic wave of expansion and recoil passes down the arteries. This is called the pulse. The pulse can be felt by placing the fingers (not thumb) on the skin at these pulse points:

- **Carotid artery** on either side of the larynx
- **Temporal artery** at the temples
- **Radial artery** at the base of thumb at the wrist

Chart 13.2

Activity	Beats per Minute Partner	Yourself
Sitting		
Using bicycle ergometer for 2 minutes		
2 minutes after completing exercise		

ACTIVITY: Observing Heart Rate

1. Count the heart rate for 30 seconds, then multiply by two.
2. Record your partner's heart rate in Chart 13.2.
3. Switch and have your partner determine your heart rate and record it. Are the rates acquired using the stethoscope and counting the pulse approximately the same? ____________________

   Explain ____________________________________

   ____
4. Count the heart rate using the pulse method immediately after performing the activities listed in Chart 13.2. Switch and have your partner determine your heart rate.
5. Record your rate and your partner's rate in Chart 13.2.
6. Compare the results obtained by your classmates. Do you discern any differences between results males and females? ______________________ ■

(text continued on page 123)

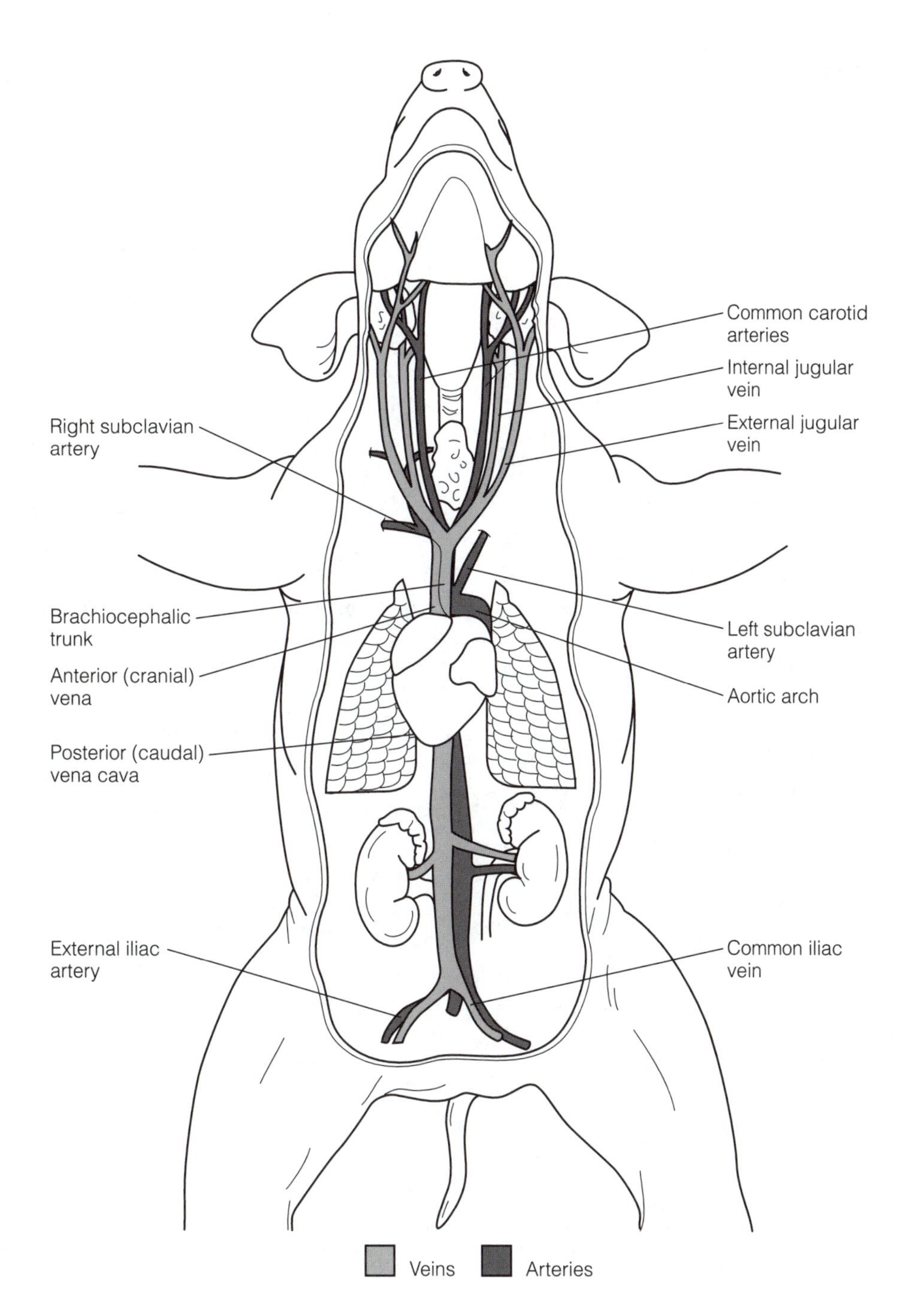

Figure 13.6 Circulatory system of the fetal pig.

Blood Pressure

Blood pressure is the force that blood exerts in all directions in any given area. It is the basis for maintaining a movement of blood from the heart through the body and back to the heart.

The pressure exerted by the blood is expressed in millimeters of mercury. The **systolic** (high) pressure is written first and the **diastolic** (low) pressure second.

The most common method of measuring the blood pressure involves the use of an instrument, the sphygmomanometer, which consists of an inflatable cuff, a rubber bulb, tubing, and a gauge.

ACTIVITY: Observing Blood Pressure

Taking Blood Pressure

To take the blood pressure:

1. Have a subject sit.
2. Fully extend the subject's arm, allowing it to rest comfortably on a table.
3. Locate the brachial artery about one inch above the elbow on the anteromedian aspect of the arm. Determine where the pulse can be felt most strongly.
4. Wrap the inflatable cuff around the upper arm slightly above this point.
5. Place the bell of the stethoscope at the pulse point.
6. Close the valve on the vent of the bulb and inflate the cuff so that the pressure reading is between 180 mm and 200 mm, which exceeds the pressure in a normal artery and stops the flow of blood in that vessel. No sound should be audible through the stethoscope at this time.
7. While listening for a pulse, release the pressure slowly by opening the valve of the bulb.
8. Record the pressure at the first sound you hear. This represents the sound of blood flowing through the artery as the result of each contraction of the ventricle (**systolic pressure**).
9. Continue lowering the pressure until there is a muffling or disappearance of sound. Record this as the **diastolic pressure.**
10. Release the remaining pressure immediately following your recording of diastolic pressure.

    Your blood pressure ________

    Your partner's ________

    Normal blood pressure varies from one individual to another. Blood pressure also varies with age and sex.
11. Compare your figures with the averages in Table 13.1.

Table 13.1 Blood Pressure Averages

Age	Sex	Systolic Pressure	Diastolic Pressure
20	female	100–130	60–85
20	male	105–140	62–80
40	female	103–150	65–92
40	male	110–150	70–94

Exercise and Blood Pressure

To determine the effects of exercise on blood pressure, do the following:

1. Take the subject's blood pressure at rest.
2. Have the subject exercise on the bicycle ergometer for two minutes. (Leave uninflated blood pressure cuff in place.) If the bicycle ergometer is not available, have the subject run in place for two minutes.
3. Repeat the blood pressure measurement *immediately* after exercise.
4. Repeat this measurement after four minutes and measure again until the pressure returns to normal.
5. Record your results below. Individuals who are on a regular exercise program may show a marked decrease in recovery time period from those who do not exercise regularly.

   At rest ________

   Immediately after exercise ________

   Time needed for blood pressure to return to normal

   ________ ■

Study Questions

Define systolic blood pressure:

____

____

____

Define diastolic blood pressure:

____

____

____

Define hypertension:

____

____

____

Define hypotension:

____

____

____

In your own words, describe the effects of hypertension and hypotension.

____

____

____

____

____

What would you expect to happen to blood pressure and pulse rate during sleep? Explain.

____

____

____

____

After exercise, what changes did you observe in systolic pressure?

____

____

____

____

In diastolic pressure?

____

____

____

____

Are these changes you expected? Explain.

____

____

____

____

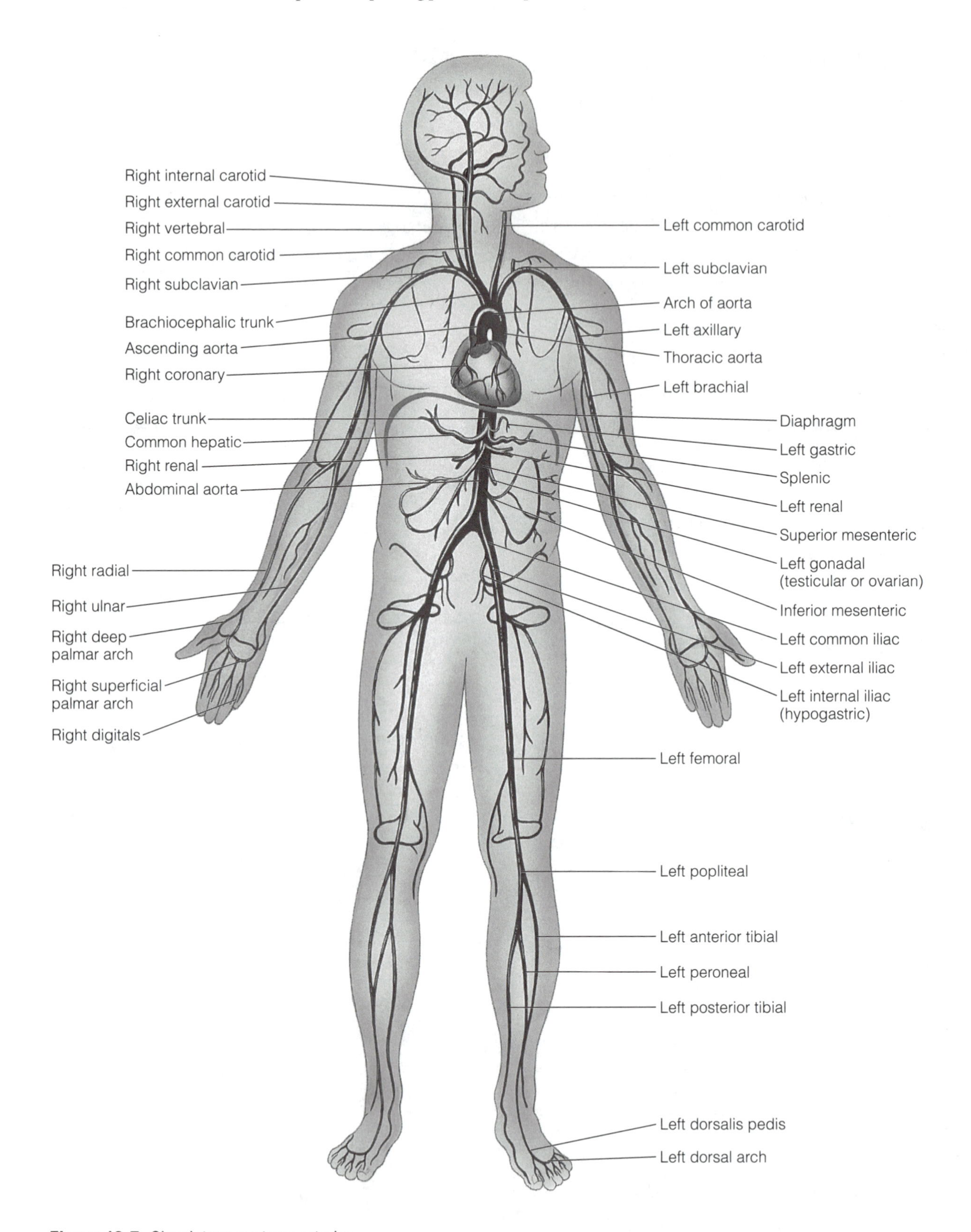

Figure 13.7 Circulatory system: arteries.

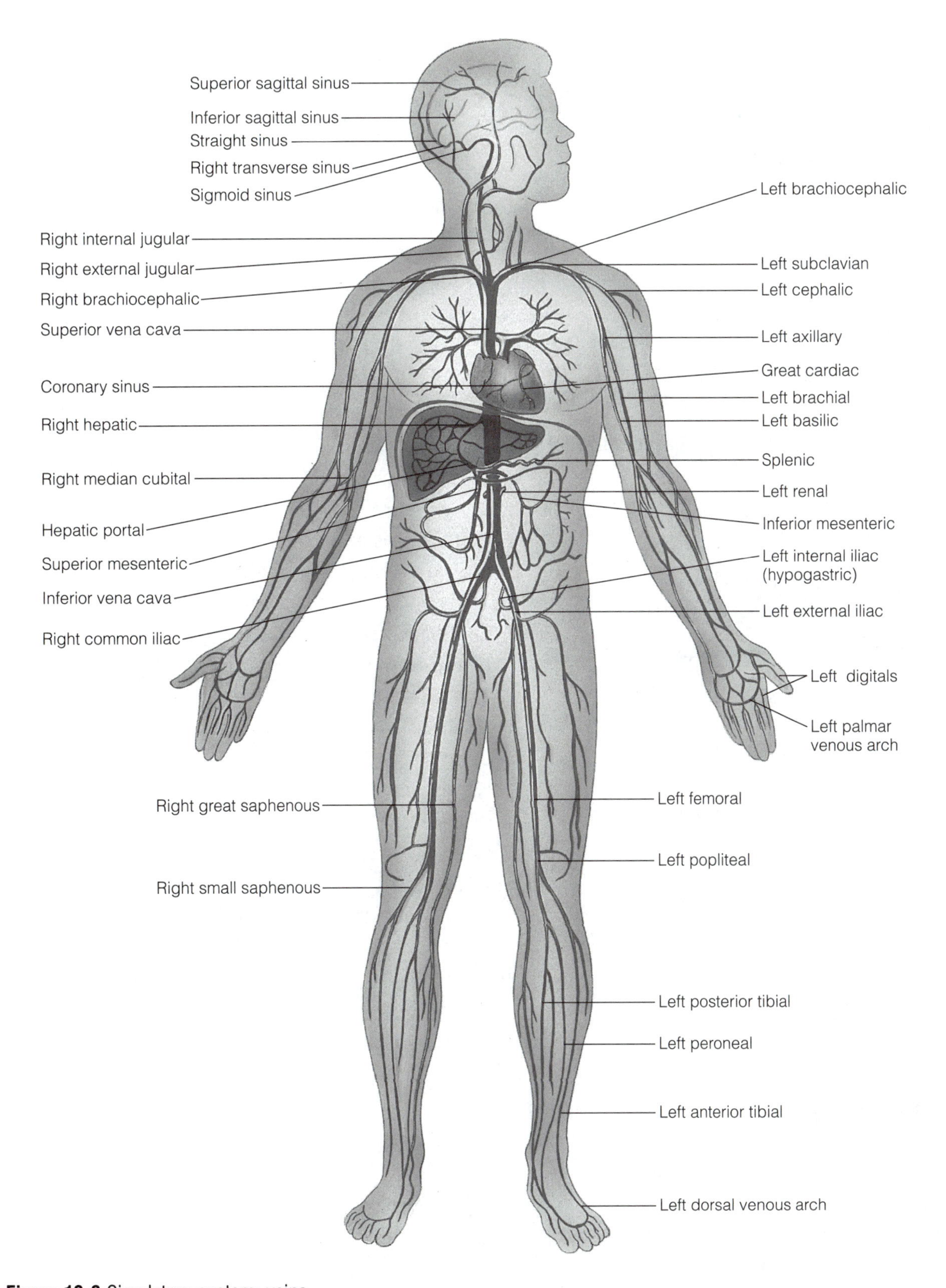

Figure 13.8 Circulatory system: veins.

EXERCISE

14 The Respiratory System

OBJECTIVES

After completing this exercise, you should be able to:

1. Locate and identify the respiratory organs and their accessory structures.
2. Understand the function of the respiratory organs and their accessory structures.
3. Demonstrate the mechanisms of lung inflation and deflation.
4. Explain the relationships between thorax size, inspiration, and expiration.
5. Demonstrate the effects of exercise on respiratory rate and composition of exhaled air.

MATERIALS

- ❑ Freeze-dried lung
- ❑ Colored pencils
- ❑ Bell jar
- ❑ Tape measure
- ❑ Two flat microscope slides
- ❑ Wet spirometer with disposable mouthpieces
- ❑ Hand mirror or glass plate
- ❑ Celsius thermometer
- ❑ 150 ml Erlenmeyer flasks
- ❑ Litmus solution in dropping bottles
- ❑ Drinking straws
- ❑ Paper sandwich bags
- ❑ Paper cups
- ❑ Sagittal section of skull

Models of:

- ❑ Sagittal section human head
- ❑ Larynx
- ❑ Lung

Introduction

Every cell of the human body requires energy to fuel the metabolic activities vital to its survival. Oxygen must be carried to the cells so that glucose can be "burned" and its energy made available. One waste by-product of the energy release process is carbon dioxide, which in large amounts is toxic to the cell.

The primary function of the respiratory system is to deliver oxygen to and remove carbon dioxide from the blood. The structures of the respiratory system transfer oxygen from the atmospheric air to the blood and transfer carbon dioxide from the blood to the exterior of the body.

The respiratory system also plays a role in maintaining a relatively constant blood pH. The changing levels of carbon dioxide (CO_2) and its solubility in water result in a buffering mechanism that prevents sudden shifts in hydrogen ion concentration.

Anatomy of the Respiratory System

In humans, the nose serves not only as an air vent but as an air conditioner. The skin within the opening of the nostrils is hairy and these hairs serve to strain out coarse particles or insects. As the air makes continued contact with the warm, moist nasal passages, it, too, becomes warmed and moistened. Moisture is added to the nasal cavity by liquid draining from four pairs of hollow areas in the bones of the skull. These spaces, called **sinuses** (Figure 14.1), are located in the frontal, ethmoid, sphenoid, and maxillary bones.

The nasal passages join the **esophagus** (food canal) just behind the mouth in the region called the throat, or **pharynx.** The union of these two passages makes it possible to breathe through the mouth. Below the throat, the air passage crosses in front of the esophagus. This makes it possible for food or drink to be sucked into the air passages, seriously blocking breathing.

Inferior to the tongue is a flap of cartilage called the **epiglottis.** During the act of swallowing, the epiglottis forms a tight seal over the air passage. This permits swallowed material to move toward the stomach and not the lung.

It is also possible for food or fluid to enter the nasal passage above. However, the **uvula**, the flap of tissue that is an extension of the palate (the roof of the mouth), closes the upper air passages. If you open your throat wide and look into a mirror, you can see the uvula. A current of air passing over the uvula may cause it to vibrate, causing snoring.

ACTIVITY: Studying the Structures of the Upper Respiratory System

1. Locate the paranasal sinuses shown in Figure 14.1 on the model of the human skull.
2. Study the structures labeled in Figure 14.2.
3. Locate the structures of Figure 14.2 on the model of the human skull.
4. Study the model of a sagittal section of the human head. ■

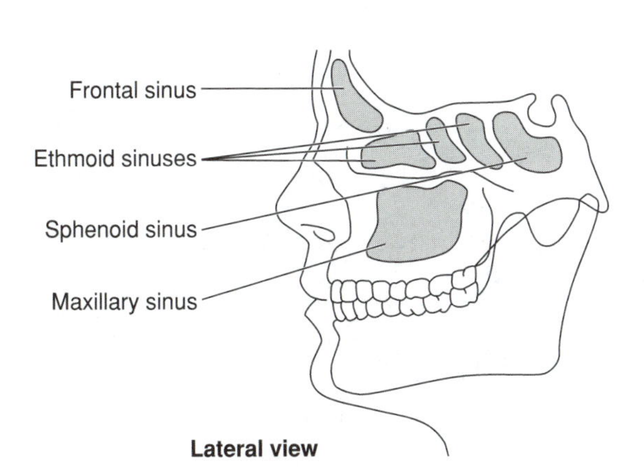

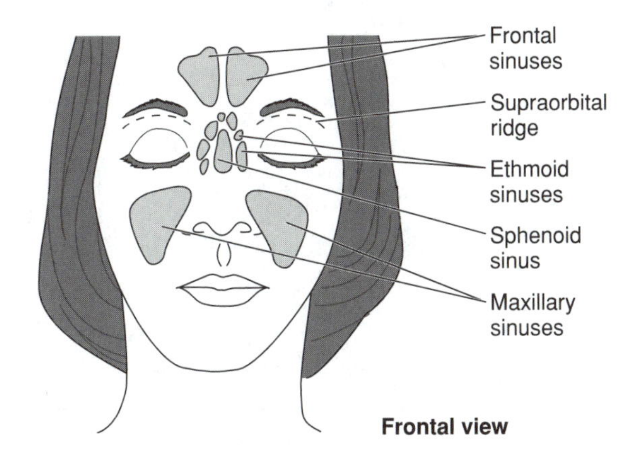

Figure 14.1 Human paranasal sinuses.

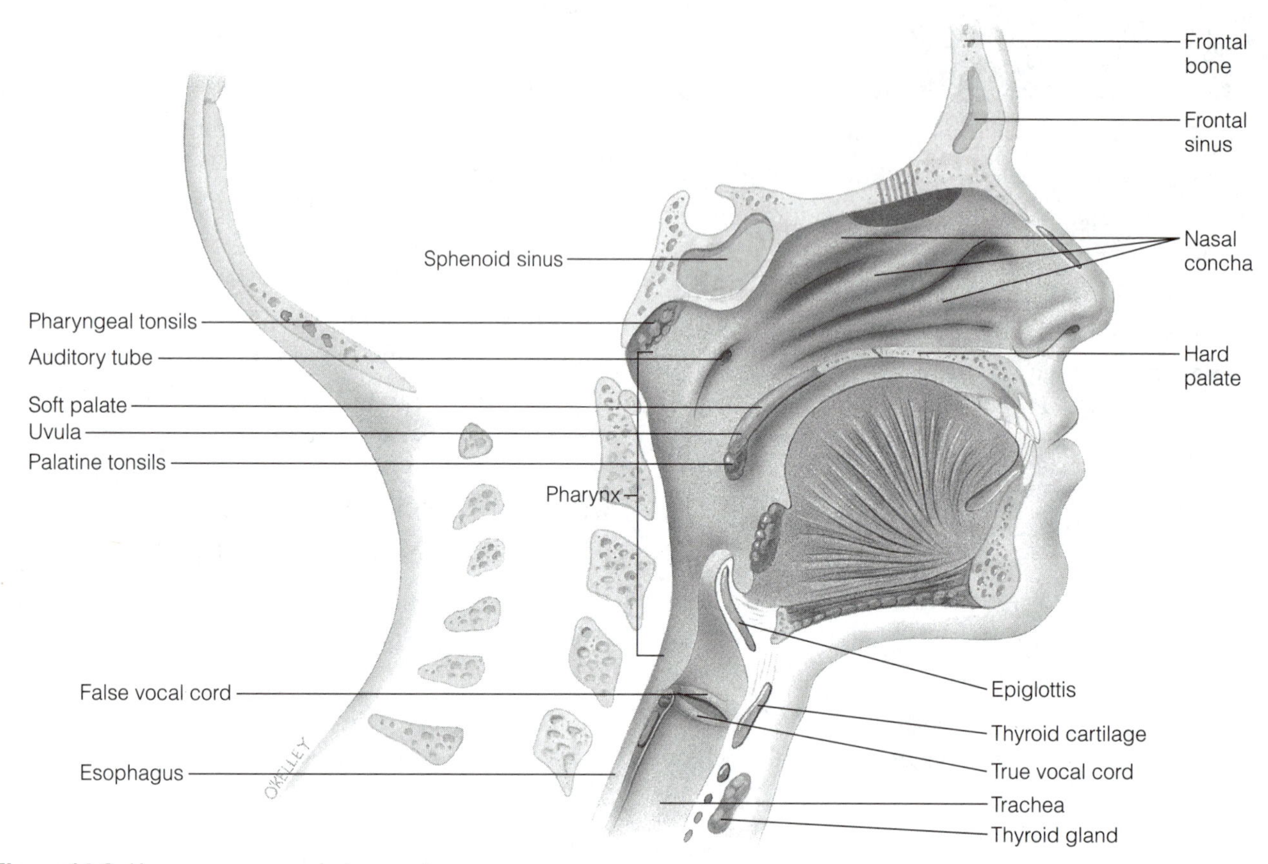

Figure 14.2 Human upper respiratory system.

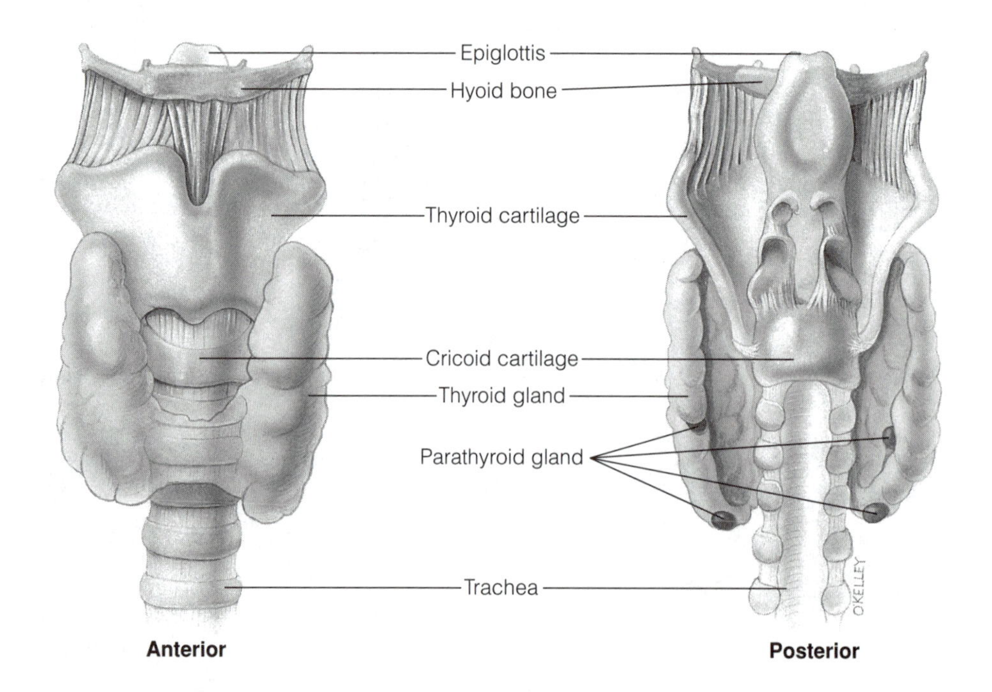

Figure 14.3 Human larynx with thyroid gland.

Larynx

In mammals, the portion of the air passage just below the epiglottis has been modified for the production of sound. The voice box, or **larynx,** is protected by a shieldlike cartilage, the **thyroid cartilage.** You can both feel and see the thyroid cartilage as the larynx moves up and down each time you swallow. The thyroid cartilage is more pronounced in men due to the influence of male sex hormones. A completely unbiblical legend claims that when Adam swallowed his bit of the apple in the Garden of Eden, he couldn't quite get it down. All his male descendants, therefore, bear the enlarged thyroid cartilage commonly called the Adam's apple.

Attached in front to the thyroid cartilage and in the back to smaller cartilages are folds of tissue called the **vocal cords.** As air is expelled past the vocal cords, they vibrate. This is clearly demonstrated when you hum—the more forceful the air, the louder the hum. It is also possible to put varying degrees of tension on the vocal cords. The more tension in the cords, the higher the pitch of the hum. In order to produce recognizable speech, one must be able to control the lips, mouth, tongue, and cheeks.

Infections of the throat and larynx change the sound of the voice, making it rough and hoarse.

ACTIVITY: Studying the Structures of the Larynx

1. Study the structures found in Figure 14.3.
2. Study the model of the larynx. Locate the structures that are labeled on Figure 14.3. ■

Respiratory Tree

Below the larynx is a tube about 4 1/2 inches long called the trachea, or windpipe. To ensure that this tube is kept open at all times, it is fitted with 16 to 20 C-shaped rings of sturdy cartilage.

A little below the point where the neck meets the trunk, the trachea divides into two branches called **bronchi**. Each bronchus leads to a separate lung. Since the bronchi subdivide and branch like a tree, it is often called the **bronchial tree** (respiratory tree). Surrounding the bronchial tree are the **lungs**. The lungs fill almost the entire chest cavity, extending from the collarbone to the thin sheet of muscle called the **diaphragm**. The right lung is slightly larger and is divided into three lobes. The left lung has two lobes. A lung that is a mere sac, like a balloon, could not deliver the amount of oxygen required by the cells. To increase their surface area, lungs divide and subdivide. They become finer and thinner until they are a cluster of microscopic **alveoli**, or air sacs. A tiny bunch of alveoli surround a thin duct leading from the bronchial tree. The oxygen molecules from the air are now at the boundary line that separates the environment from the tissue. How the oxygen is carried to the cells of the body was discussed in Exercise 12.

It is estimated that the lungs contain some 600,000,000 alveoli. As oxygen passes from the lung to the tissue, carbon dioxide (CO_2), a waste product of energy production, takes the reverse path. The air we exhale is lower in oxygen content and higher in carbon dioxide than the air we inhale.

ACTIVITY: Studying the Structure of the Respiratory Tree

1. Study the structures labeled in Figure 14.4.
2. Color the trachea blue, the bronchi green, the epiglottis yellow, and the larynx brown.
3. Study a model of the lungs. ■

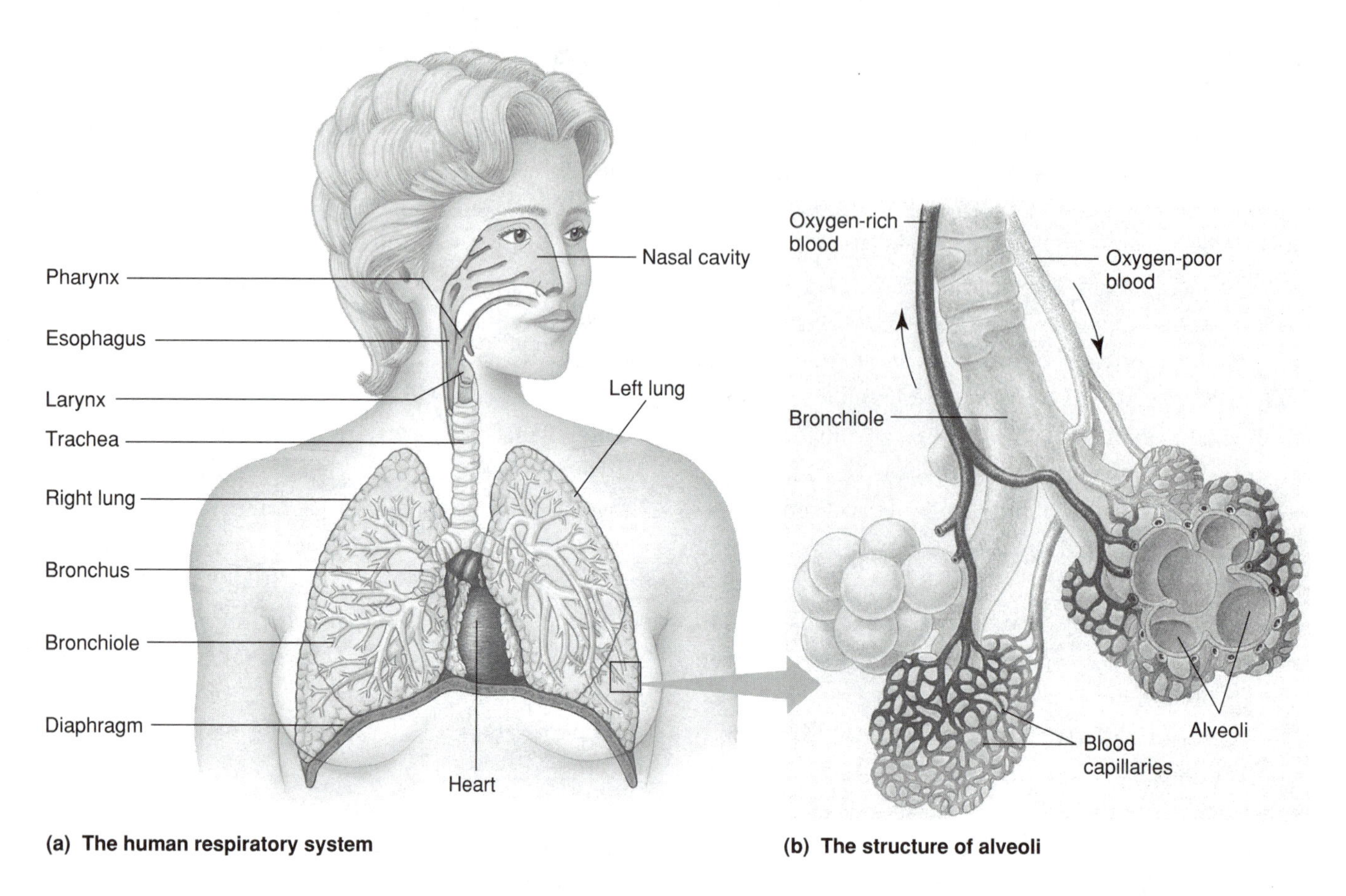

Figure 14.4 Human lung, trachea, and nasal passages.

Mechanics of Respiration

The air within the lung is not sufficient to maintain life for more than a few minutes. We renew the oxygen by breathing 14 to 20 times a minute, moving old air out and drawing new air into that remaining in the respiratory passageways.

The pleura are continuous membranes surrounding each lung, somewhat like a fist thrust deep into the surface of an inflated balloon (Figure 14.5).

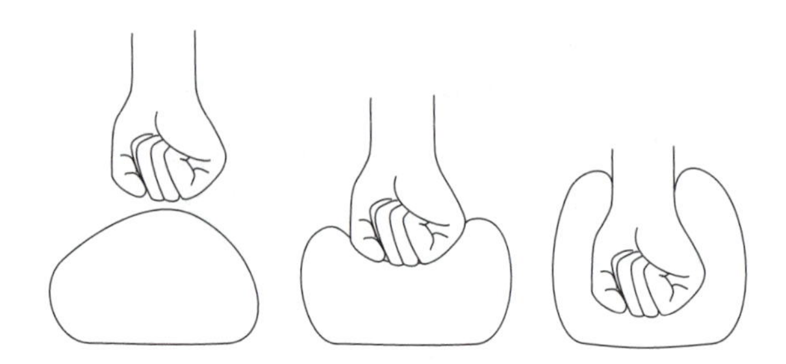

Figure 14.5 Mechanics of the pleura.

How do we move air into and out of the lungs? Lung tissue is elastic and stretches. Its normal tendency is to recoil. Each lung is surrounded by a membrane called the **pleura**, which adjoins the lung, then doubles on itself and lines the chest cavity. Fluid between the two layers of the pleura permits the lungs to stretch and recoil with minimum friction. There is also a partial vacuum between the two pleural layers, which causes the lungs to stick to the chest wall as it expands.

As the diaphragm contracts and flattens and the external intercostal muscles contract and lift the ribcage, the pressure drops within the thoracic cavity. Air begins to rush in through the respiratory tract and fill the lungs. At the same time, the adhesion between the lungs and the inside of the expanding thoracic cavity helps to expand the lungs. With both the lungs and the thoracic cavity expanding, air races in to fill the space (volume) created. Muscles called the scalenes and the sternocleidomastoids can also help to raise the ribcage and further expand the thoracic cavity. This is the process of **inspiration**.

During **expiration**, the diaphragm ceases to receive stimulation and intra-abdominal pressure pushes the diaphragm up to its relaxed "dome-shaped" position. Internal intercostal muscles and gravity help to drop the ribcage and thoracic cavity back to its smaller, more compressed size. These thoracic changes and the natural recoil of the lungs increases pressure inside the lungs and forces air out.

ACTIVITY: Movement of the Diaphragm During Respiration

1. Gently palpate your xiphoid process (the most inferior portion of the sternum). This is the anterior attachment of the diaphragm.
2. Stand erect and place your hands over your upper abdomen.
3. Inhale deeply, while keeping your abdominal muscles relaxed.
4. Notice how your abdomen is pushed outward as the contracted diaphragm flattens.

   Describe the position of the contracted diaphragm.

   ____

   ____

   ____

   ____

   Describe the position of the relaxed diaphragm.

   ____

   ____

   ____

   ____

   How does the position of the diaphragm affect the size of the thoracic cavity?

   ____

   ____

   ____

   ______________________________ ■

ACTIVITY: Circumference of the Chest During Respiration

1. Working with a partner, use a tape measure to illustrate that changes in chest size do, in fact, play a role in the movement of air.
2. Write your results in Chart 14.1.
3. While standing erect, measure the circumference of the chest three inches below the collarbone. Be certain that each measurement is made at the same level of the chest. ■

Chart 14.1 Measurement of Chest Circumference

Activity	Chest Measurement: Inches	Chest Measurement: Centimeters
at end of quiet inspiration		
at end of quiet expiration		
at end of forced inspiration		
at end of forced expiration		

ACTIVITY: Changing Pressures and the Movement of Air

Gases move from an area of greater pressure to one of less pressure. If equal volumes of gas are placed in containers of different sizes, the gas in the larger container will be under less pressure.

1. Obtain a lung-thorax model (Bell jar with balloons). Complete Chart 14.2.

   What could you do to the lung-thorax model to cause the balloons to fill with air?

   ____

   ____

   ____

   ____

   Try your hypothesis using the lung-thorax model. What happens to the Bell jar compartment and the balloons as you manipulate the rubber sheeting?

   ____

   ____

   ____

   ____________________ ■

Chart 14.2

Bell Jar	Human Body Equivalent
straight glass tubing	
glass wall of jar	
rubber sheeting	
balloons	
Y-shaped glass tube	

ACTIVITY: The Pleura and Pulmonary Volume

Recall the location of the double-layered pleura with their intervening serous fluid. As the wall of the thorax with its adhering pleura layers enlarges, the lungs covered by a pleura layer follow the expansion of the chest wall.

1. Obtain two dry, flat microscope slides (they will represent the double-layered pleura).
2. Place one slide directly on top of the other.
3. Slide the slides back and forth over each other.
4. Lift one slide straight up off the lower slide. Do the slides slide easily over each other? Do the slides separate easily? __________
5. Wet the surface of one slide and again sandwich the two slides together.
6. Slide the slides over each other.
7. Attempt to separate the two slides by lifting up or prying them apart. Do the slides slide easily over each other? __________

   Do the slides move more easily, less easily, or the same as when there was no intervening fluid?

   ____

   Are you able to easily pry the two slides apart?

   ____

   If the integrity of the pleura is disturbed (such as by a puncture into the chest wall), what do you think would happen to the lung?

   ____

   ____

   ____

   ____________________ ■

Respiratory Volumes

It is possible to alter the amount (volume) of air you breathe and also how rapidly you breathe in and out. This makes it possible for you to vary your oxygen consumption according to the demands of your body.

The amount of air an individual can move in one normal inhalation and exhalation is called **tidal volume**. **Inspiratory reserve volume** is the amount of air that can be forcibly inhaled after a normal inhalation. **Expiratory reserve volume** is the amount of air that can be forced out of the lungs beyond tidal volume. The maximum amount of air an individual can forcibly inhale and exhale is the **vital capacity**. Even after forcible exhalation, there still remains over one quart of air; this is the **residual volume**.

ACTIVITY: Observing Respiratory Volumes

The spirometer is an instrument used to measure the volumes of air in the lung. Before making any measurements, familiarize yourself with the instrument. *Do not inhale* through the disposable mouthpiece. Exhale through the mouthpiece, holding your nose. Read the exhaled volume from the scale on the instrument. The indicator needle will remain in position until you reset it.

Work in pairs.

Determining Tidal Volume

1. Set the indicator needle at 0.
2. Inhale normally through your nose.
3. Exhale a normal breath through the mouthpiece while pinching the nose closed. Inhale normally through the nose.
4. Repeat ten times without resetting the spirometer. Divide by 10 to determine your tidal volume.
5. Repeat three times.
6. Record your data in Chart 14.3.

Expiratory Reserve Volume

1. Place the indicator needle at 0.
2. Inhale and exhale a normal breath through your nose.
3. Then, without inhaling again, blow as much air as you can through the spirometer mouthpiece while pinching the nose closed.
4. Reset the indicator needle at 0 and repeat twice more.
5. Record your data in Chart 14.3.

Vital Capacity

1. Set the indicator at 0.
2. Take two or three deep breaths, inhaling as much air as possible.
3. Exhale steadily through the mouthpiece until you can force no more air from your lungs. Remember to pinch your nose closed.
4. Reset the indicator needle at 0, and repeat twice more.
5. Record your data in Chart 14.3. ■

Chart 14.3

Test Performed	1	2	3	Average	Normals
Tidal volume					300–500 ml
Expiratory reserve					1000–1500 ml
Vital capacity					4000–4800 ml

ACTIVITY: The Temperature and Composition of Exhaled Air

Temperature of Exhaled Air

The respiratory system also has an excretory function. Heat and moisture as well as carbon dioxide leave the body in exhaled air.

1. Exhale through your mouth onto a glass plate or mirror. Your results indicate that one component of exhaled air is ________________.
2. Record the temperature of the room air. ________°C.
3. Breathe on the bulb of a Celsius thermometer several times. Record the temperature. ________°C.

   What might you conclude causes the change in temperature?

   ____

   ____

   Think: If the room temperature recorded had been quite warm, say 40°C, what would be the temperature of your expired air?

   ____

   ____

   ____

Composition of Exhaled Air

Carbon Dioxide

The carbon dioxide from expired air will form an acid in a water solution.

CO_2	$+ H_2O$	$\rightarrow H_2CO_3$	$\rightarrow H^+ +$	HCO_3^-
Carbon dioxide	Water	Carbonic acid	Hydrogen ion	Bicarbonate ion

The acid-base indicator can, therefore, estimate the presence of carbon dioxide exhaled into water. You will use such an indicator in the following experiment.

Carbon dioxide is a waste product formed during the cellular respiration process. It is increased after exercise.

To prove that exhaled air contains carbon dioxide, working in pairs, perform the following experiment.

1. Place 30 ml of water in each of two 150 ml flasks. Add 10 drops of litmus solution to each. Litmus is an indicator which is blue when basic, pink when acidic. Carbon dioxide combines with water to form an acid which changes blue litmus to pink.
2. Using a straw, bubble air (from the lungs) into one solution. Record the time it takes for the solution to turn pink.
3. Exercise vigorously for 2 minutes and repeat the experiment using the second flask. Record data in Chart 14.4.

Chart 14.4

	Time for Color Change
Quiet expiration	
Following exercise	

Hydrogen Ions

Hydrogen ions are the primary controlling factor in the rate and depth of respiration. When breath is held, carbon dioxide increases, thereby increasing the concentration of hydrogen ion. (Review the formula for the reaction of carbon dioxide with water.)

Determine the time that a subject can hold his or her breath in each of the following situations:

1. Breathe normally several times, then hold the breath as long as possible.
2. Hyperventilate for 30 seconds, then exhale and hold the breath.
3. Rebreathe into a paper bag for 60 seconds, then hold breath once again, then attempt to breathe. *Terminate the experiment if the subject shows any signs of distress.* Record data in Chart 14.5.

Chart 14.5

Type of Breathing	Minutes Breath Held
Normal	
Hyperventilation	
Rebreathing in bag	

If you sit quietly, listening to a lecture for instance, your breathing becomes shallow. Carbon dioxide levels increase and you feel the urge to ventilate the lungs. The result? A yawn. Unfortunately, the lecturer recognizes the yawn as a sign of sleepiness and boredom (which it is) rather than an attempt to break the shallow breathing rhythm that leads to sleep (which it also is). The open yawn and/or falling asleep during a lecture is unacceptable to some professors. ■

ACTIVITY: Breathing Patterns

1. Work in pairs.
2. Seat your lab partner comfortably and place yourself so that you may observe chest movements and nasal and oral inhalations without staring face to face, which would make your subject uncomfortable and aware of his or her breathing.
3. Ask the subject to perform the following activities. After each activity, record the results in Chart 14.6. Count respirations for 30 seconds. At the same time, observe any deep inhalations or remarkable breath holding. Allow normal breathing to resume between each test.
   a. Normal breathing
   b. Drinking 8 ounces of water quickly
   c. Reading aloud for 1–2 minutes
   d. Counting (not aloud) backwards from 500 to 450
   e. Exercising vigorously for 3 minutes

Chart 14.6

Activity	Respirations/Minute	Depth of Respiration	Number of Deep Inhalations
Normal breathing			
Drinking			
Speaking			
Mental activity			
After exercise			

Which activity decreases respiratory rate?

___

___

Which activity increases respiratory rate?

___

___

Can you usually speak aloud and inhale simultaneously?

___

___

Can you swallow and breathe simultaneously?

___

___

Propose a reason why prolonged sedentary mental activity might produce a yawn. (Boredom is not a valid response.)

___

___ ■

Deglutition (Swallowing) Apnea

When you swallow, impulses traveling from the tongue, pharynx, and larynx to the respiratory center of the medulla inhibit inspiration. This is a defense mechanism that helps to prevent aspiration of liquid and solid food into the trachea. You will note how this reflex affects the impulse to inspire.

ACTIVITY: Inhibition of Inhalation

1. Hold your breath until the urge to breathe becomes overwhelming.
2. Do not inspire but take a small sip of water. What has happened to your urgent need to breathe? ■

Study Questions

1. Name the anatomical structure located directly behind the trachea.

___

2. Name the membrane that covers the lungs.

___

3. What is the name for the structure that blocks entry into the larynx during swallowing?

___

4. What is the name given to the groups of air sacs that comprise most of the lung tissue?

___

5. What is the Adam's apple?

___

___

6. Would it be possible for a healthy child to die by holding the breath?

___

Why?

___

___

___

7. Briefly describe the effect of carbon dioxide on breathing.

___

___

___

___

8. What effect does exercise have on carbon dioxide production?

____

____

9. What structures of the respiratory system are involved in the following?

   a. Humidifying air

   ____

   ____

   b. Warming air

   ____

   ____

   c. Filtering air

   ____

   ____

   d. Conducting air

   ____

   ____

   e. Producing sound

   ____

   ____

10. What is the function of the diaphragm?

____

____

____

____

11. What activities can influence breathing?

____

____

____

____

12. Differentiate cellular respiration and ventilation. (Use a reference text.)

____

____

____

____

____

____

____

____

13. During inhalation, is air sucked in or pushed in?

____

Explain:

____

____

____

____

____

____

14. Does your chest expand because your lungs inflate, or do your lungs inflate because your chest expands?

____

____

Explain:

____

____

____

____

____

____

15. What structural feature ensures that the trachea will be an open airway?

____

____

EXERCISE

The Digestive System

OBJECTIVES	MATERIALS
After completing this laboratory exercise, you should be able to: 1. Identify the main structures of the digestive system in the fetal pig and human models. 2. List and describe the basic functions of the main digestive system structures. 3. Describe the role of enzymes in chemical digestion. 4. Describe the role of each of the enzymes of the digestive system.	❑ Fetal pigs ❑ Human torso models ❑ Heat-resistant test tubes (10 ml and 40 ml) ❑ Test tube holders ❑ 20 ml graduated cylinder ❑ Deionized water ❑ Glucose solution (2%) ❑ Starch solution (1%) ❑ Amylase solution (1%) ❑ Benedict's solution ❑ Boiling water bath ❑ Peanuts and stand ❑ Celsius thermometer ❑ Matches ❑ Wax pencil or china marker ❑ Dissecting equipment

Introduction

Animals cannot make their own food as plants do. Therefore, they need to obtain, from other organisms, molecules that serve as energy sources and building materials. The large molecules contained in the food we eat must first be broken down into their basic building blocks before they can be absorbed by the digestive tract. This process of digestion is eventually accomplished in two ways: **mechanical digestion**, which means that larger food particles are physically broken up into smaller pieces, and **chemical digestion**, which produces the chemical change from macromolecules (protein, starches, or lipids) to their smaller building blocks (amino acids, sugars, fatty acids, and glycerol). Proteins, called enzymes, usually help with chemical digestion (Table 15.1).

Table 15.1 Major Digestive Enzymes and Chemical Reactions

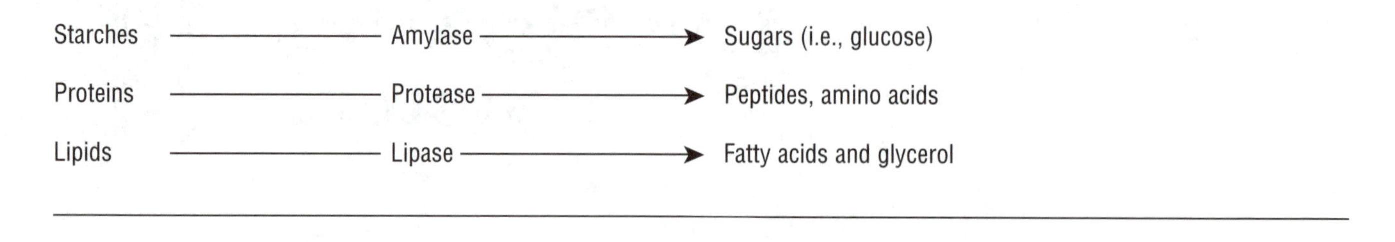

Substrate	Enzyme	Products
Starches	Amylase	Sugars (i.e., glucose)
Proteins	Protease	Peptides, amino acids
Lipids	Lipase	Fatty acids and glycerol

Give an example of a place in the body where mechanical digestion is performed.

Give an example of a place in the body where chemical digestion is performed.

Digestive System Structures

Refer to Figure 15.1 when studying the structures of the digestive system.

Eating and Swallowing

The **oral cavity**, or mouth, is a place where both mechanical digestion (teeth and tongue) and chemical digestion (amylase from salivary glands) occurs. The amylase produced by salivary glands is carried to the mouth by ducts and breaks starches down into sugars. The **pharynx**, or throat, which actually begins behind the nasal cavity, is where swallowing starts. The **esophagus** is the muscular tube where swallowing continues until food reaches the stomach.

The Stomach

Chemical digestion occurs in the stomach, as does mechanical digestion via crushing and churning of food from an extra layer of smooth muscle. The stomach produces both a protease (pronounced pro-tee-ase) called pepsin and hydrochloric acid (HCl). Pepsin, which becomes active only in acidic environments, breaks large proteins down into smaller subunits. HCl activates pepsin and is the source of the stomach's acidity. Although not an enzyme, HCl's corrosive properties permit it to break up connective tissues for easier digestion.

Consider that the food you eat and the air you filter as you breathe are not sterile. What other benefit might HCl provide in the stomach?

Which category of food molecule (protein, starch, or lipid) has not been broken down significantly by enzymes by the time food has left the stomach?

The Small Intestine

The small intestine, particularly its first section (the duodenum), is the major site of chemical digestion. A lipase, amylase, and two proteases are all carried from where they are made in the pancreas to the duodenum. The pancreas also contributes sodium bicarbonate to the duodenum in order to neutralize the stomach's HCl. Another duct brings bile for mechanical digestion of fats from the liver and gallbladder. Bile is not an enzyme and does not break any chemical bonds in lipids. As a **surfactant** (similar to detergents), bile does break large fat globules into smaller droplets (a mechanical change), allowing the lipase to work more efficiently. The duodenum is an important focal point for chemical digestion, but it is relatively short (10 inches). Another 20 feet of small intestine follows the duodenum. This allows plenty of time and distance for chemical digestion by all of those enzymes added to the duodenum. The first half of the

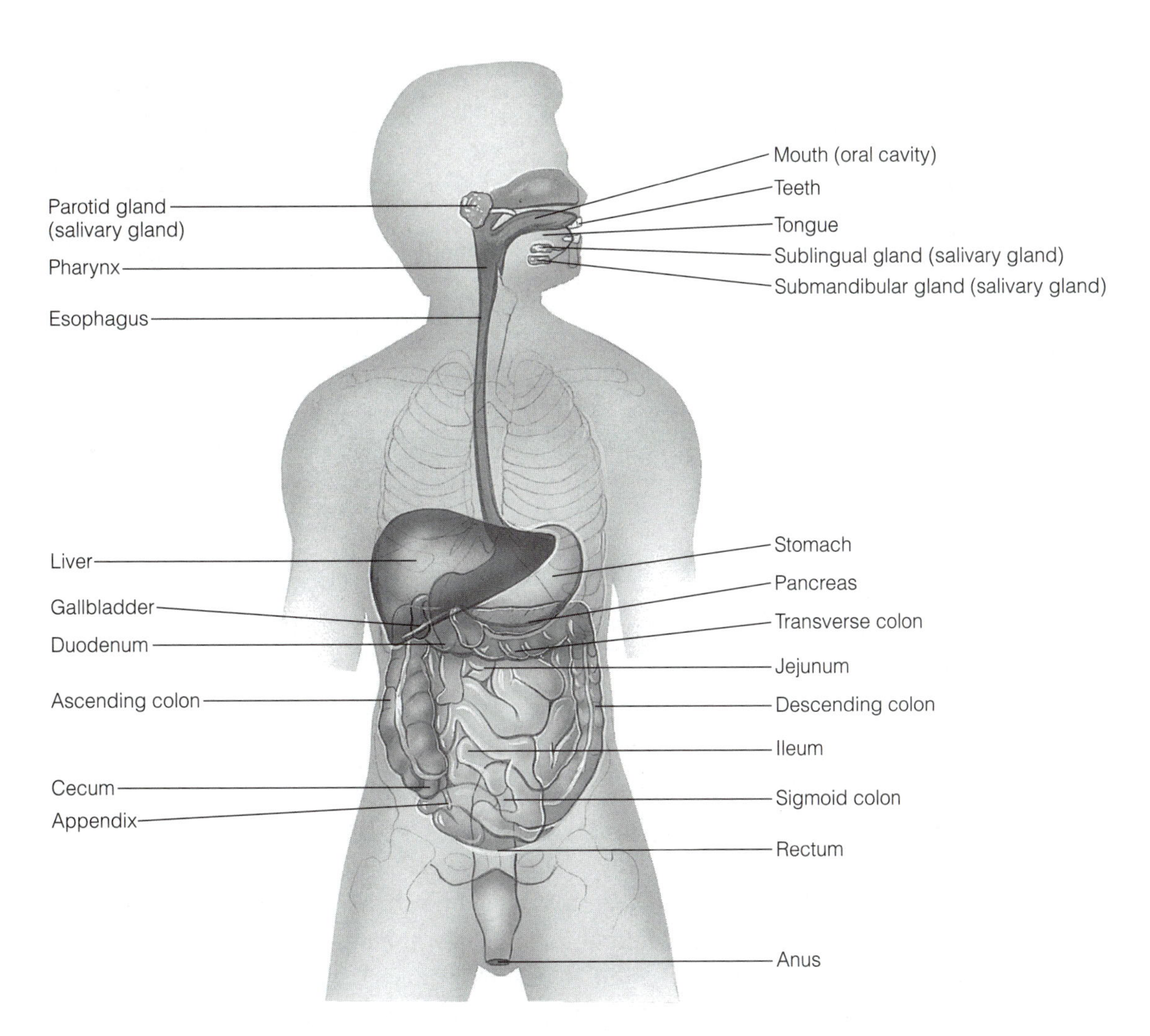

Figure 15.1 The human digestive system.

small intestine is the **jejunum**, and the second half is called the **ileum**. "Finishing" enzymes are found farther down the small intestine, and break down nearly digested molecules to their smallest, easiest-to-absorb size.

The small intestine, particularly the latter half, is also the site of most nutrient and water absorption (Figure 15.2). For this reason, both the inner wall and the columnar cells of the small intestine have a highly folded appearance. These small fingerlike projections of the intestinal wall are called **villi**, and the submicroscopic folds in the cell membrane are called **microvilli**. Both structural modifications serve to increase surface area for absorption.

What two "nonenzymes" are contributed to the duodenum, and what are their jobs?

____

____

____

____

Would a protease produced by the pancreas or a protease produced by the stomach work best in an acidic environment?

____

____

In an alkaline environment?

____

____

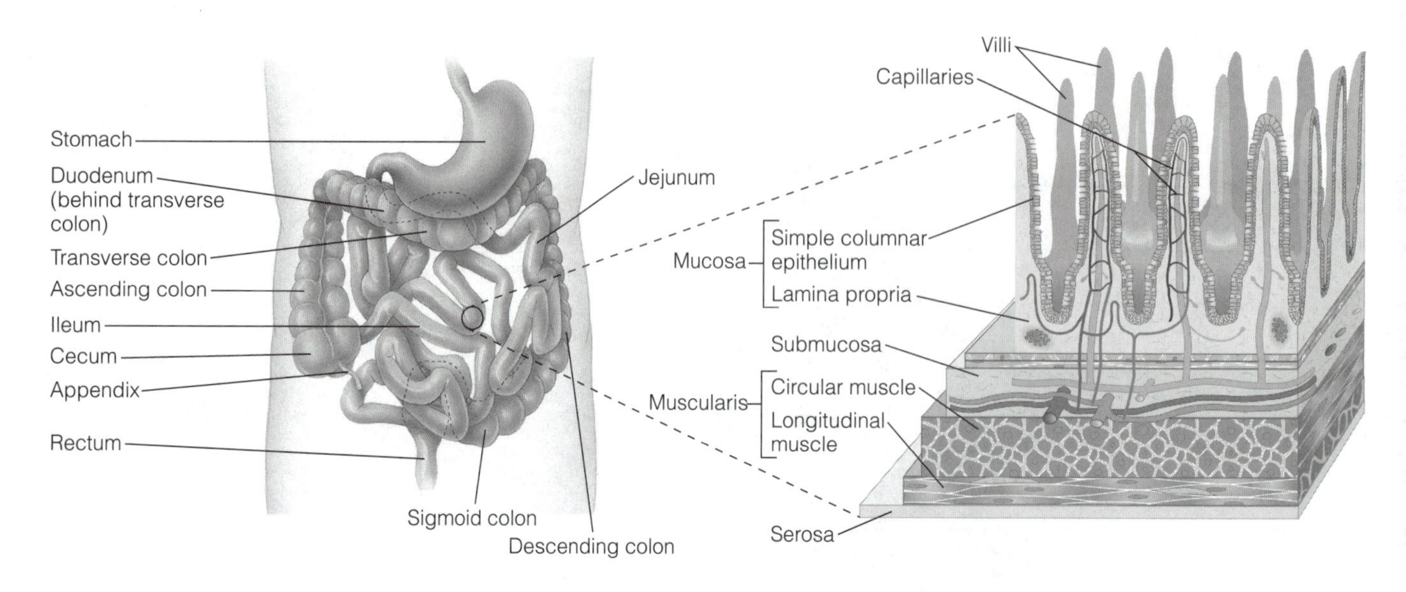

Figure 15.2 Wall of the digestive tract.

The Large Intestine

The large intestine is mainly involved in water absorption and solidification of the digestive wastes. Nearly all of the digestion and absorption that is possible should have been completed by the time material enters the large intestine. Microorganisms that live in the large intestine break down some of these wastes and actually provide some vitamins such as vitamin K.

What parts of the digestive system are not involved with either digestion or absorption of food?

____

____

____

____

What are their functions?

____

____

____

____

Digestive Tract Tissues

The digestive tract is a tube composed of four basic tissue layers. These layers are modified for each specialized part of the digestive system. The outer layer, the **serosa**, is an epithelial membrane. The **muscularis** is the thickest part of the intestinal wall and is made of smooth muscle. Two different directional layers of this muscle allow for **peristalsis**, or the wavelike propulsion of material through the digestive tract. The next inner layer, the **submucosa**, is a loose connective tissue bed for holding the large blood and lymphatic vessels and nerves that travel along the digestive tract. The **mucosa** is the innermost layer, which is composed of simple columnar epithelium and its underlying loose connective tissue. This loose connective tissue bed, the **lamina propria**, holds capillaries needed for absorption.

ACTIVITY: Anatomy of the Digestive System

Identify the following structures on the human models and also during your fetal pig dissection. Use Figures 15.3a and 15.3b for reference.

- **Liver** large, multilobed structure under lungs and diaphragm.

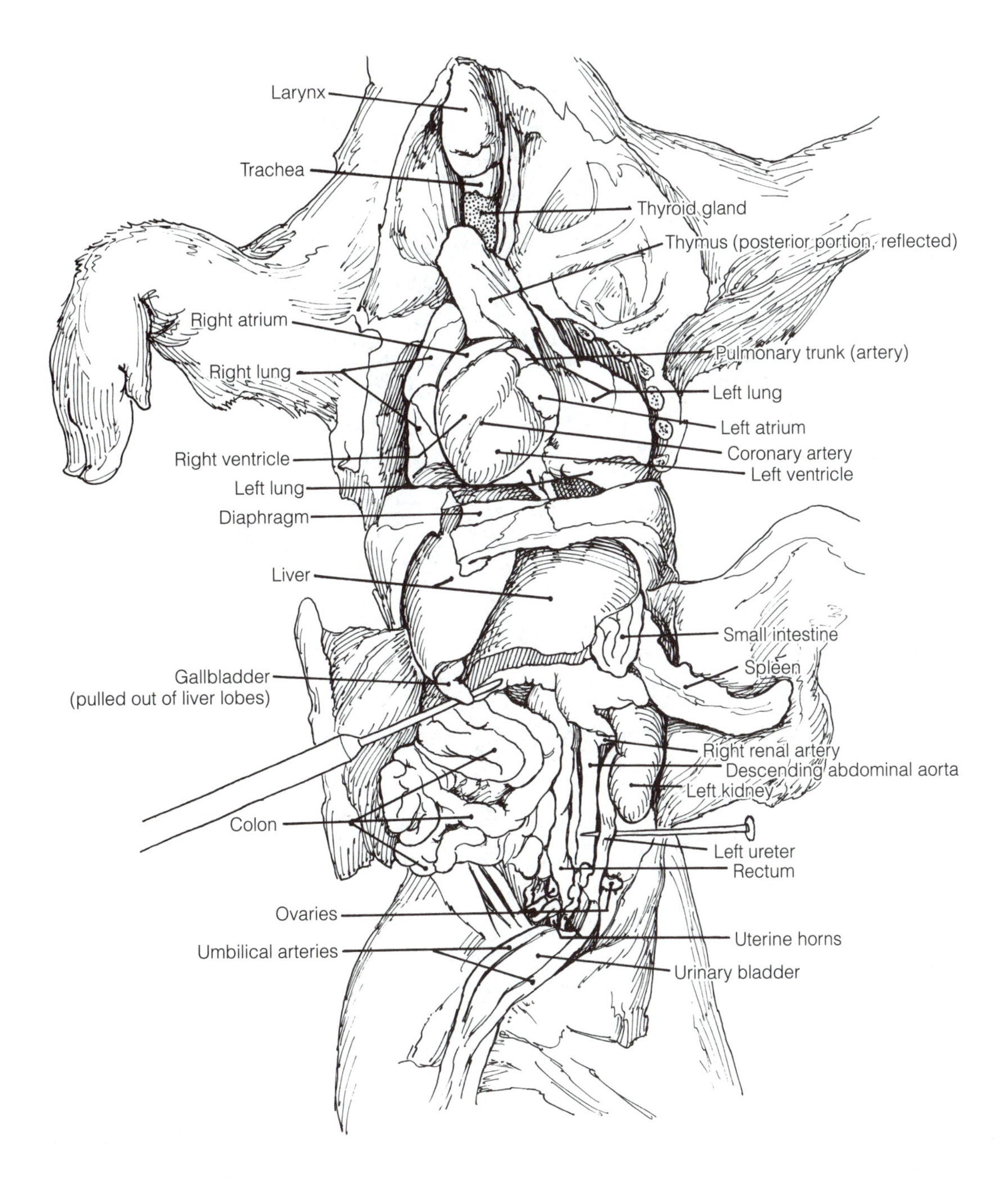

Figure 15.3a Fetal pig digestive system.

- **Gallbladder** lift up the lobes of the liver on the right side to find the small, green, saclike organ.
- **Esophagus** lies directly behind the trachea; follow its path from its beginning in the neck to its entry into the stomach.
- **Stomach** saclike organ under the liver on the left side.
- **Duodenum** first part of the small intestine (2–3 inches in the fetal pig) attached to the right side of the stomach.
- **Pancreas** begins in the loop between the stomach and duodenum and ends near the spleen on the left side; looks like a clump of about a hundred pinhead-sized grapes.
- **Jejuno-ileum** the coiled remainder of the small intestine in the fetal pig.

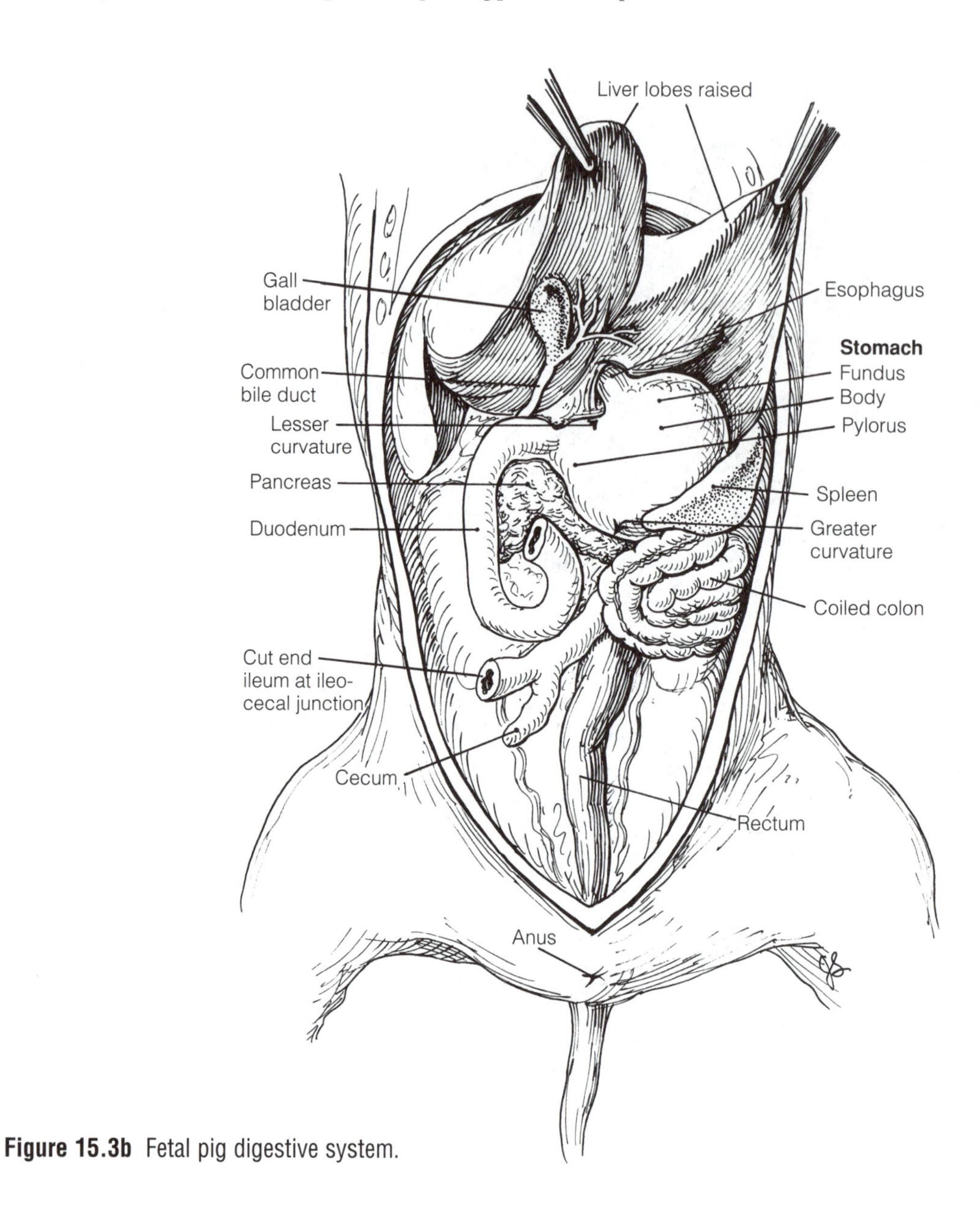

Figure 15.3b Fetal pig digestive system.

- **Cecum** the thumblike blind pouch at the junction of the jejuno-ileum and colon. This is where the appendix is located in humans.
- **Colon** the usually darker, thicker, coiled (large) intestine in the fetal pig.
- **Rectum** the last one to two inches of large intestine leading to the anus.

Compare the approximate length and diameter of the small and large intestine. Which is longer?

____

Which is larger in diameter?

____

Why is the small intestine so named?

____

____

____

______________________________________ ■

Metabolism

A **calorie** is a measure of **heat energy** (the amount of energy needed to raise 1 ml of water 1° Celsius). Because most of our metabolic efforts go toward maintaining body temperature, it is not surprising that this unit of measure is used by nutritionists to quantify the amount of energy stored in food.

The calorie is a very small unit of measure compared to the size of the human body and the energy it uses, so nutritionists are actually using **kilocalories,** (1 kilocalorie = 1000 calories) to describe the energy content of food. To simplify, the term **Calorie** (with a capital C) is used here to describe kilocalories.

Just as in factory warehouse management, if more material enters than is shipped out, inventory accumulates. Likewise, if we consume more Calories than we use, the remainder is stored. Specifically, if a portion of the chemical energy that was taken in is not used, it is mostly converted into a more permanent and lightweight molecular form (adipose tissue).

The following activity is intended to demonstrate the connection between calories and heat energy, and also that the human cell is much more efficient at extracting a peanut's energy than fire would be.

ACTIVITY: The Peanut Test

1. Use a graduated cylinder to measure 25 ml of cold tap water, and pour into a 50 ml Kimax or Pyrex (heat-resistant) test tube.

2. Place a Celsius thermometer into the test tube and read the temperature after 2 full minutes. Record ________°C.

3. Attach a whole peanut to the end of the wire in the metal dish (don't worry if this takes more than one attempt). If a peanut set-up is not available, a large pin pushed through a cork or styrofoam base from beneath may be substituted (wet the cork or styrofoam to reduce flammability).

4. Once the peanut is securely attached, light a match and hold it under the peanut until the peanut will burn on its own. (This may take 5 seconds or longer.)

5. Using a test tube holder, hold the test tube containing the 25 ml of water and the thermometer over the flame. *Note:* Do not rest the thermometer at the bottom of the test tube! Raise it off the bottom and use it to *gently* stir the water.

6. If the peanut burns out after a few seconds, relight it and try again. The peanut normally burns for 1 to 3 minutes.

7. Once the peanut is burned out, record the temperature of the water in degrees Celsius. ________°C. Subtract your starting temperature (from step #2) from this temperature and record the rise in temperature. ________°C.

8. Recall that a small calorie (not a kilocalorie) is the amount of heat required to raise one ml of water 1°C. Since the amount of water in your experimental test tube was 25 ml, you must multiply the rise in temperature you obtained by 25 to determine how many calories were transferred from the burning peanut to the water.

   Formula: Temp. change X 25 (ml) = calories.

   ________°C X 25 = ________ calories.

9. Also recall that the Calories described by nutrition sources are kilocalories. To convert from calories to Calories, simply divide your answer in #8 above by 1000.

   Formula: calories/1000 = ________ Calories

10. Suppose a cup of peanuts is estimated to 250 peanuts, and that this cup of peanuts should have a Calorie count of about 850 Calories. Multiply your answer from step #9 by 250 and record.

    ________ Calories.

    Compare the 850-Calorie estimate with your experimental results for a cup of peanuts. Is there any significant difference between the two numbers?

    ____

    If so, how much higher or lower?

    ____

Give two reasons for any possible difference between your experimental results and the estimate given by a diet book.

Given that 2000 Calories per day is an average energy requirement for a person, how many peanuts per day would be needed to sustain life (if they were the lone food source)?

ACTIVITY: Enzyme Function

The purpose of this activity is to demonstrate the role of enzymes in the digestion of large food molecules. You will do this activity in groups of four.

1. Label five small test tubes 1–5 with a wax pencil. Add the following to each tube:
   #1 5 ml deionized water and 5 drops of Benedict's solution
   #2 5 ml glucose solution and 5 drops of Benedict's solution
   #3 5 ml warm starch solution and 5 drops of Benedict's solution
   #4 5 ml amylase solution and 5 drops of Benedict's solution
   #5 5 ml warm starch solution, 2 ml amylase solution, and 5 drops of Benedict's solution
2. Place in a test tube rack for 10 minutes, occasionally shaking the tube gently. Next, heat all tubes for 5 minutes in a boiling water bath. Remove with test tube holders and place in the rack.
3. A color change from blue to either green, orange, or yellow is considered a positive test for the presence of sugar.

Which test tubes contained sugar at the start of the experiment?

Which additional tube(s) tested positive for sugar at the end of the experiment?

Where did this sugar in the "new" test tube come from?

What chemical reaction has occurred in the tube you answered for Question #2?

What is the purpose of the other tubes?

What is an enzyme?

Give an example of an enzymatic reaction occurring in humans that relates to this experiment.

Study Questions

1. List, in order, the major organs of the digestive tract from beginning to end.

2. Explain the difference between mechanical and chemical digestion.

3. List the three main "macromolecules" described in this exercise and the "building block molecules" each is made of.

4. What is peristalsis, and what tissue layer of the digestive tract causes it to occur?

5. In which part of the digestive tract does most digestion occur?

   What "accessory glands" assist the digestive process?

6. In which part of the digestive tract does most absorption take place?

   How are the lining cells of this structure modified to facilitate absorption?

7. List the three classes of enzymes and the molecules they help break down.

8. In which part of the digestive system does the most significant amount of fat digestion occur and why?

9. What is bile?

10. In what ways would the anatomy of the human and the fetal pig digestive systems be similar?

    How are they different?

EXERCISE

16 The Urinary System

OBJECTIVES

After completing this exercise, you should be able to:

1. List and locate the major structures of the urinary system.
2. Understand the functions of these structures.
3. Relate the microscopic structure of the kidney to the gross anatomy of each structure.
4. Understand three major processes involved in kidney function (filtration, reabsorption, and tubular secretion).

MATERIALS

- ❑ Kidney stone samples (demo)
- ❑ Sheep kidneys, triple-injected (2/table)
- ❑ Fetal pig
- ❑ Combistix test papers for pH, protein, and glucose (demo table)
- ❑ Unknown urine samples A, B, and C
- ❑ Human torso
- ❑ Human kidney models
- ❑ Urine hydrometers and hand magnifier
- ❑ Demonstration: cross section of the kidney

Urinary System

Excretion is the elimination of cellular chemical wastes from the body (metabolic wastes). Some of the major cellular waste products are CO_2, H_2O, urea, and salts.

Human skin is an excretory organ, eliminating H_2O, salts, and urea via perspiration. As we exhale, the lungs excrete CO_2 and H_2O. The large intestine excretes salts and some water. The major excretory organs are the kidneys, which filter the blood and eliminate urea, salts, and water as urine.

Proteins eaten by humans are digested in the stomach and small intestine and enter the blood as amino acids. In the liver, amino acids are de-aminated. This produces the poisonous compound ammonia (NH_3) which is converted to the less toxic urea. Urea, water, and salts are eventually removed from the blood by the kidneys. Following is the equation for urea formation in the liver:

$$CO_2 + 2\,NH_3 \longrightarrow \underset{\text{urea}}{(NH_2)_2C{=}O} + H_2O$$

The Kidneys

The human kidneys are two flattened, bean-shaped organs about the size of a clenched fist. They are located in the dorsal part of the abdominal cavity behind the peritoneum, lateral to the second lumbar vertebra. The right kidney is lower than the left (Figure 16.1).

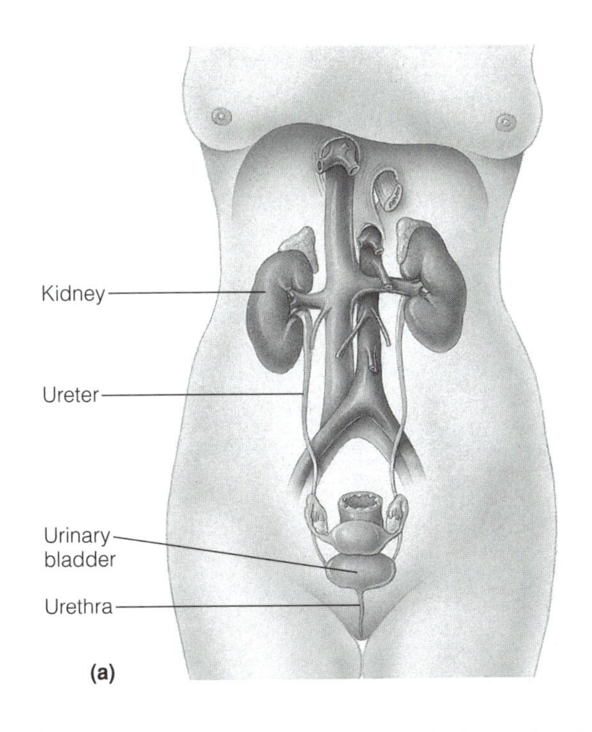

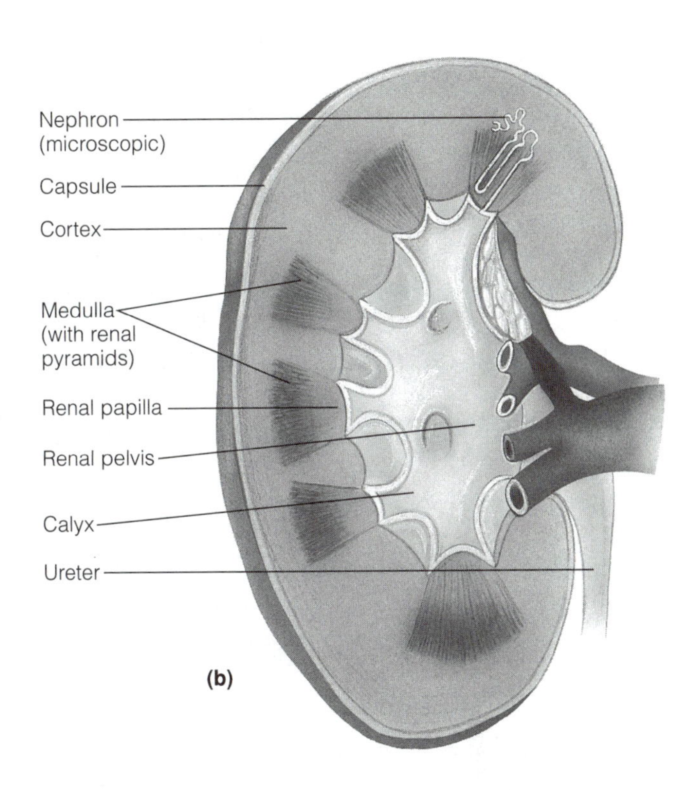

Figure 16.1 The kidney: a) location and b) internal anatomy.

Although there is some disagreement on this fine point of anatomy, the kidneys may technically be considered as *retroperitoneal* rather than classified as being part of the abdominal cavity. Retroperitoneal means behind the abdominal cavity lining, and part of the posterior body wall.

Gross Anatomy of the Kidneys

The **capsule** is the protective fibrous outer covering. The renal artery and renal vein punch through the capsule on the "puckered" medial side of each kidney.

Inside, the kidney has two distinct areas. The **cortex** is the outer portion of the organ, and has a granular texture. The vast majority of the capillary blood flow is found in the cortex, which is composed of the renal corpuscles (the glomerulus and surrounding capsule), the proximal convoluted tubules and distal convoluted tubules.

Although the term **medulla** generally refers to the inner portion of an organ, the renal medulla appears as the middle portion of the kidney, and is composed of the **pyramids**. The renal or medullary pyramids are typically ten or more triangular striated structures between the cortex and a series of tubes leading out of the kidney. Within the pyramids are **loops of Henle** and **collecting tubules (ducts)**. The **papillae** are the pointed tips of the pyramids where they insert into the opening of a tube called the **calyx**. Filtered urine is carried to the **pelvis**, which is the innermost hollow muscular part of the kidney. The pelvis serves to receive filtered, processed urine which will be emptied into the ureters and carried to the bladder for temporary storage.

The Nephron

Each kidney contains about one million microscopic units called **nephrons**, which collectively provide much surface area to filter wastes from the blood. The nephron consists of a **glomerulus, glomerular (Bowman's) capsule, proximal convoluted tubule, loop of Henle, distal convoluted tubule,** and **collecting tubule**. Each portion of the nephron has a specific role to play in filtration, reabsorption, and secretion of wastes. Refer to Figure 16.2 as you continue reading the following explanation.

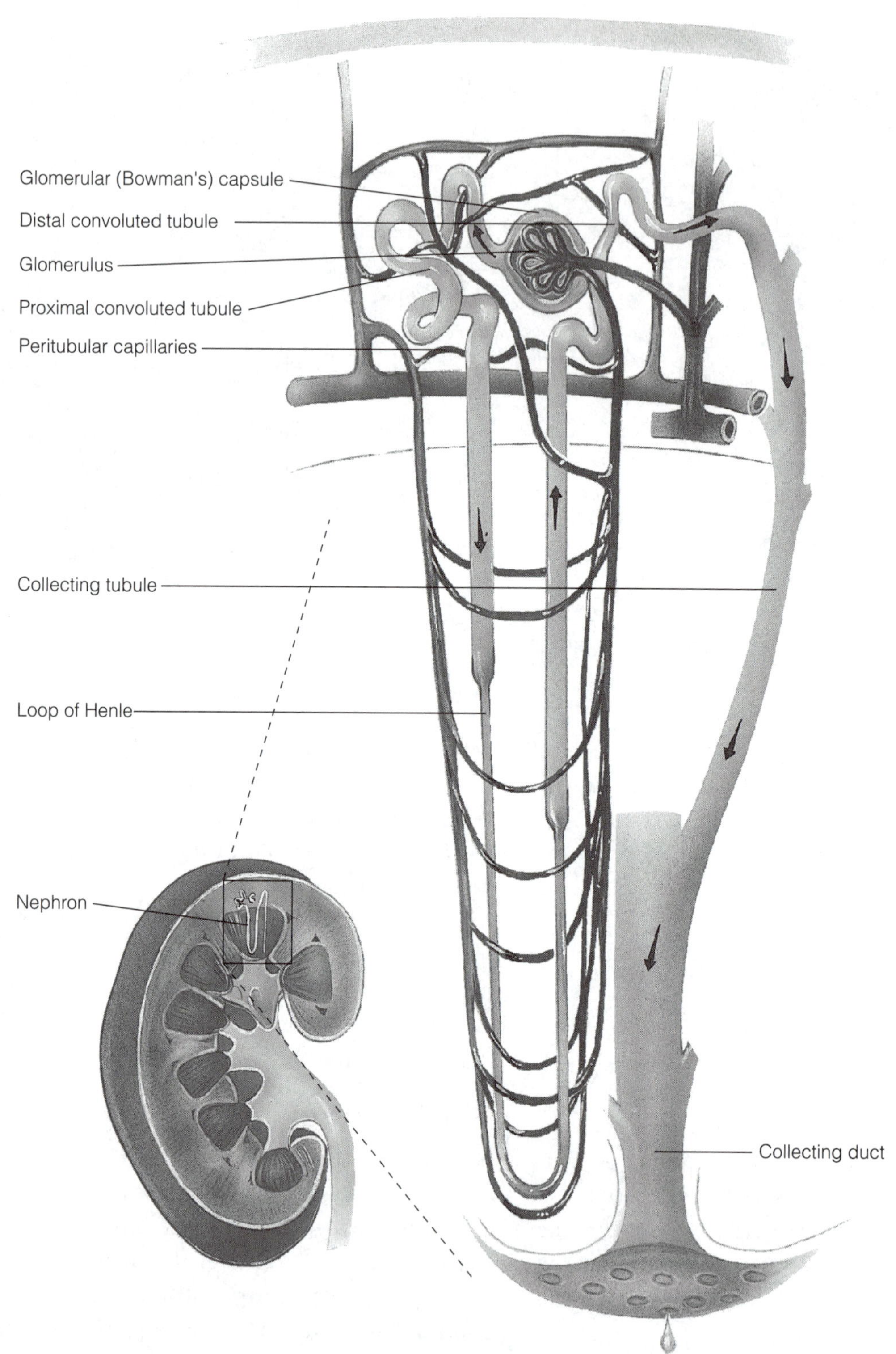

Figure 16.2 Location and drawing of the nephron.

The Glomerulus and Glomerular (Bowman's) Capsule

Blood containing excretory wastes enters the **interlobular artery**, which branches into several **afferent arterioles.** Each afferent arteriole enters an "arterial knot," or **glomerulus**, a microscopic capillary network that sits within **glomular (Bowman's) capsule**, a membranous cuplike structure. High blood pressure in the glomerulus causes **filtration** to occur from the glomerulus into the glomular capsule. The filtered material at this point is called **glomerular filtrate.** It contains urea, amino acids, vitamins, hormones, salts, glucose, ions, and water—but no large proteins or blood cells. Every day the kidneys produce about 180 quarts of glomerular filtrate, but most of the nutrients, vitamins, hormones, and water are reabsorbed, so the average amount of urine is only about 1 pint to 1 quart per day.

Proximal Convoluted Tubule

Reabsorption occurs mainly in the walls of the proximal convoluted tubule, the first in a series of microscopic tubules. Here most of the glucose, amino acids, hormones, ions, vitamins, and some of the salts re-enter the blood by *facilitated* and active transport. After these have been reabsorbed, most of the water is then *passively* reabsorbed due to the change in concentration of solutes in the proximal tubule.

Loop of Henle

The loop of Henle arises from the convoluted tubule and is composed of a descending limb, a hairpin loop, and an ascending limb. Located in the medulla, it functions for additional reabsorption and secretion. It also plays a major role in making the medulla a hypertonic environment for fluid passing through in the collecting tubules and ducts.

Distal Convoluted Tubule

The last portion of the renal tubule, the distal convoluted tubule, is located in the renal cortex. It is active in reabsorption and secretion of ions, including H^+ (hydrogen ions).

Collecting Tubule (Duct)

The collecting tubule or duct receives filtrate from several distal convoluted tubules and empties into the calyces and pelvis. As fluid moves through the collecting tubule or duct, some reabsorption and secretion, particularly of ions and urea, may occur. Regulation of water reabsorption occurs here. The collecting duct is not particularly permeable to water, but can be made very water-permeable when water reabsorbtion is necessary.

An important pituitary hormone, **anti-diuretic hormone** (ADH), regulates the reabsorption of water in the kidney tubules. When we exhale, and to some extent when we perspire, we lower the water level of our blood, thus increasing the salt concentration. This apparent increase in the salt concentration of our body fluids triggers both the hypothalamic thirst center, as well as the release of ADH. ADH flows from the pituitary gland, causing more reabsorption of water by the collecting tubules, thus conserving water.

Diuretics are substances that increase the flow of urine. They include alcohol, caffeine, and "water pills" taken to reduce edema.

Another hormone, **aldosterone**, secreted by the adrenal glands, promotes sodium ion and water reabsorption, as well as potassium secretion. As already mentioned in our study of the endocrine system, aldosterone is released in response to low sodium, high potassium, and low blood pressure.

Another process called **tubular excretion** (or simply secretion) also occurs. Through tubular excretion, the cells in the proximal and distal tubules pass materials into the urine. Some of the secreted chemicals include a nitrogenous compound called creatinine, and certain ions (H^+ and K^+) and drugs.

Why do you think drugs must be administered continually during infection?

____

____

____

____

The final product of filtration, reabsorption, and tubular excretion is **urine**, which leaves the distal tubules and enters collecting tubules. The collecting tubules eventually empty into the kidney pelvis (or renal pelvis). From the kidney pelvis the urine travels through a ureter to the urinary bladder, and finally exits the body to the environment via the urethra. The urethra is longer in males because it passes through the penis and carries sperm discharges as well as urine to the environment. In females, the urethra is shorter and is not part of the reproductive system (see Figure 16.1).

Some urinary disorders are:

- **Incontinence** loss of voluntary control of urination due to severed spinal cord, aging, stroke, or other factors.
- **Urinary calculi** commonly known as "kidney stones," stonelike crystals of calcium phosphate, calcium oxalate, amino acids, or uric acid. These crystals collect in the kidney pelvis and ureter. They can cause painful urination and serious kidney damage necessitating surgery, and in some cases, removal.
- **Cystitis** inflammation of the urinary bladder.

ACTIVITY: Studying the Structures of the Urinary System

Identify the structures mentioned in the previous discussion on the torso models and in the fetal pig.

Fetal Pig Dissection

During this dissection, care must be taken in order not to destroy parts of the reproductive system associated with the urinary tract (Figure 16.3).

1. Locate the paired, bean-shaped kidneys lying against the dorsal wall of the abdomen.
2. Free one kidney without severing its blood vessels or ducts, and note that it is covered over with a thin layer of peritoneum on its ventral surface only.
3. Find the long urinary duct, or ureter, extending posteriorly from the hilus, or concave depression of the kidney.
4. Trace the course of this duct to its point of entrance into the urinary bladder. This bladder empties posteriorly into a broad tube, the urethra.
5. Postpone locating the urethra until you investigate the reproductive system.

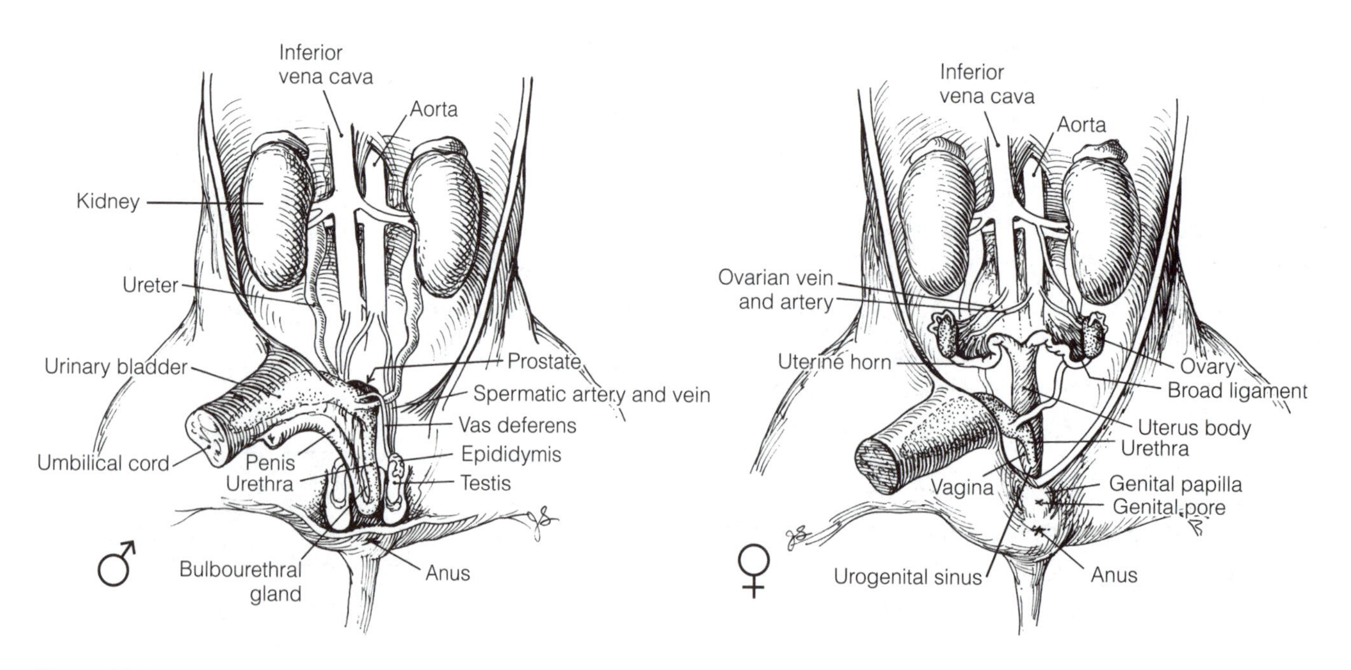

Figure 16.3 Male and female pig urinary systems.

Sheep Kidney

1. Observe the preserved triple-injected sheep kidneys at your lab table.

2. Using Figures 16.1 and 16.2 on pages 149 and 150 as a guide, locate the three main regions of the kidney:
   - **Cortex** the outermost region, which contains the glomeruli and Bowman's capsules.
   - **Medulla** the middle region, where the loops of Henle and collecting ducts form the pyramids.
   - **Kidney (renal) pelvis** the inner "hollow cavity" region.

3. If they are still attached to the kidney, locate the renal vein, renal artery, and ureter. Place the kidney in a culture dish and use the stereoscope to try to locate glomeruli, Bowman's capsules, proximal and distal tubules, and loops of Henle.

Model of the Human Kidney

Observe the models (normal and diseased) of the human kidney on demonstration.

Urine Hydrometer

1. Observe the three urine hydrometers on demonstration. Since buoyancy is related to the amount of solute in water, the higher the "float" rises, the more solute the water contains. The specific gravity is this measurement of buoyancy/ solute compared to pure water.

2. Read the specific gravity of each urine sample from the scale on the stem of the hydrometer.

3. Record your readings in Chart 16.1.

Chart 16.1

Sample	Specific Gravity
A	
B	
C	

What is specific gravity?

____

____

____

____

____

What do the differences in specific gravity in the three urine samples mean?

____

____

____

____

____

For a diabetic, would you expect a high or low specific gravity? Why?

____

____

____

____

____

Combistix Test

Using Combistix test papers, test each of the unknown urine samples labeled D, E, and F on the demonstration table for the presence of glucose, protein, and pH.

1. Remove one strip of test paper from the container and close it immediately.
2. Dip the test portion into the sample to be tested.
3. Compare the color of the paper with the color charts shown on the "Combistix" bottle.
4. Record your results in Chart 16.2.

   Which is the normal urine specimen? Why?

   ____

   ____

   ____

   ____

   Is there a single normal urine pH?

   ____

What could cause glucose to appear in urine?

____

____

____

____

What could cause protein to appear in urine?

____

____

____

____

Kidney Stones

Observe samples of human kidney stones on the demonstration table.

What might be responsible for kidney stone formation?

____

____ ■

Chart 16.2 Results of Unknown Urine Sample Tests

Sample	pH	Glucose	Protein
D			
E			
F			

Study Questions

1. What is the first product of filtration in the human nephron called?

   ____

   ____

   What is the final excretory fluid called?

   ____

2. Define excretion.

   ____

   ____

   ____

   ____

3. Name three major processes involved in removing wastes from human blood via the nephrons.

   a. ______________________________

   b. ______________________________

   c. ______________________________

4. The three major constituents of urine are:

   a. ______________________________

   b. ______________________________

   c. ______________________________

5. Name at least six components of glomerular filtrate.

   ____

   ____

   ____

   ____

   ____

   ____

6. Name all of the nephron parts in proper order, starting with the glomular (Bowman's) capsule and ending with the tubes that empty into the kidney pelvis.

   ____

   ____

   ____

   ____

   ____

   ____

7. Starting with the afferent arteriole, name in proper order the flow of blood within the nephron until it re-enters the interlobular vein. (Use Figure 16.2 for reference.)

   ____

   ____

   ____

   ____

   ____

   ____

8. Trace the flow of urine from the kidney pelvis to the environment.

   ____

   ____

   ____

   ____

   ____

   ____

EXERCISE

17 The Male Reproductive System

OBJECTIVES

After completing this exercise, you should be able to:

1. List and locate major structures of the male reproductive system.
2. Understand functions of these structures and the pathway of materials through them.
3. Relate microscopic structure of testes to gross anatomy.

MATERIALS

❑ Human torso model—male reproductive system

Prepared slides:

- ❑ Human testes—cross-section (2 per 4 students)
- ❑ Human sperm (2 per 4 students)

Introduction

The biological function of the reproductive system is unique in that its purpose is to ensure the survival of the species. The other organ systems of the body function primarily to maintain the life of the individual.

Anatomy of the Male Reproductive System

The male gonad consists of two oval-shaped **testes**, which produce the male reproductive cells or **sperm**. Each testis is enclosed in a **scrotal sac**, an outpocketing of the body wall. This external location of the testes serves an important thermoregulation function. Since moderately warm temperatures can kill sperm, this arrangement radiates heat away from the testes so that their temperature is 1–3°C cooler than the rest of the body.

Each testis contains 500-800 microscopic **seminiferous tubules,** which produce sperm by meiosis (reduction division). The process of sperm formation is called **spermatogenesis**. Sperm contain only one-half the chromosome number of the parent (23) as do the ova from the female. Fertilization restores the 46 chromosomes per cell. See Figures 17.1 and 17.2.

Wedged between the seminiferous tubules are **interstitial cells**, which produce the male hormone **testosterone**. This hormone is responsible for male secondary sex characteristics, including development of the beard, deeper voice, axillary hair, body hair and pubic hair, and development of the penis. It promotes muscle development; therefore, male bodies usually appear more muscular than do females. In other vertebrate animals, testosterone is also responsible for male secondary sex traits such as antlers, horns, bird plumage color, comb and wattle, tail feather development, and spurs.

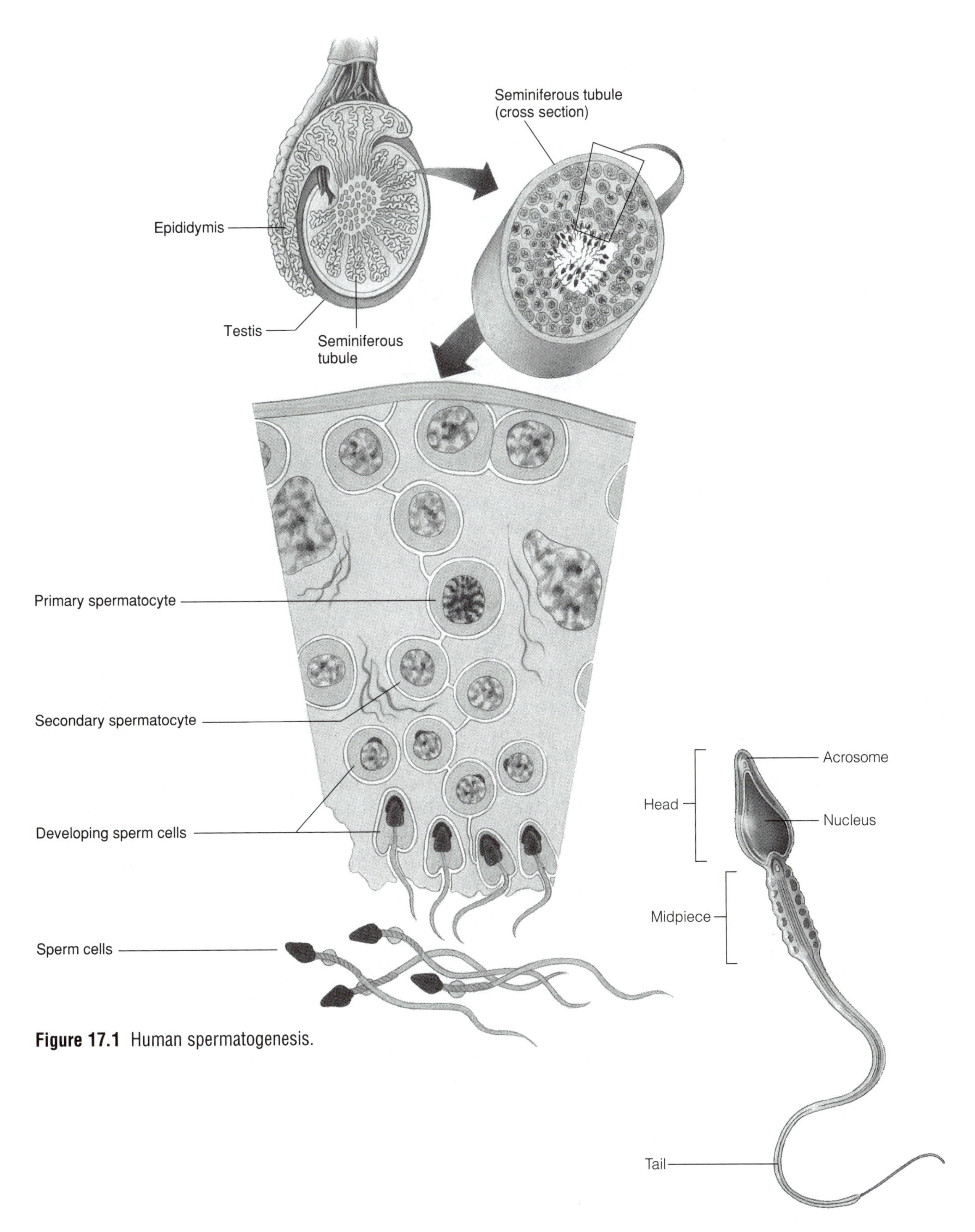

Figure 17.1 Human spermatogenesis.

Figure 17.2 Human spermatazoan.

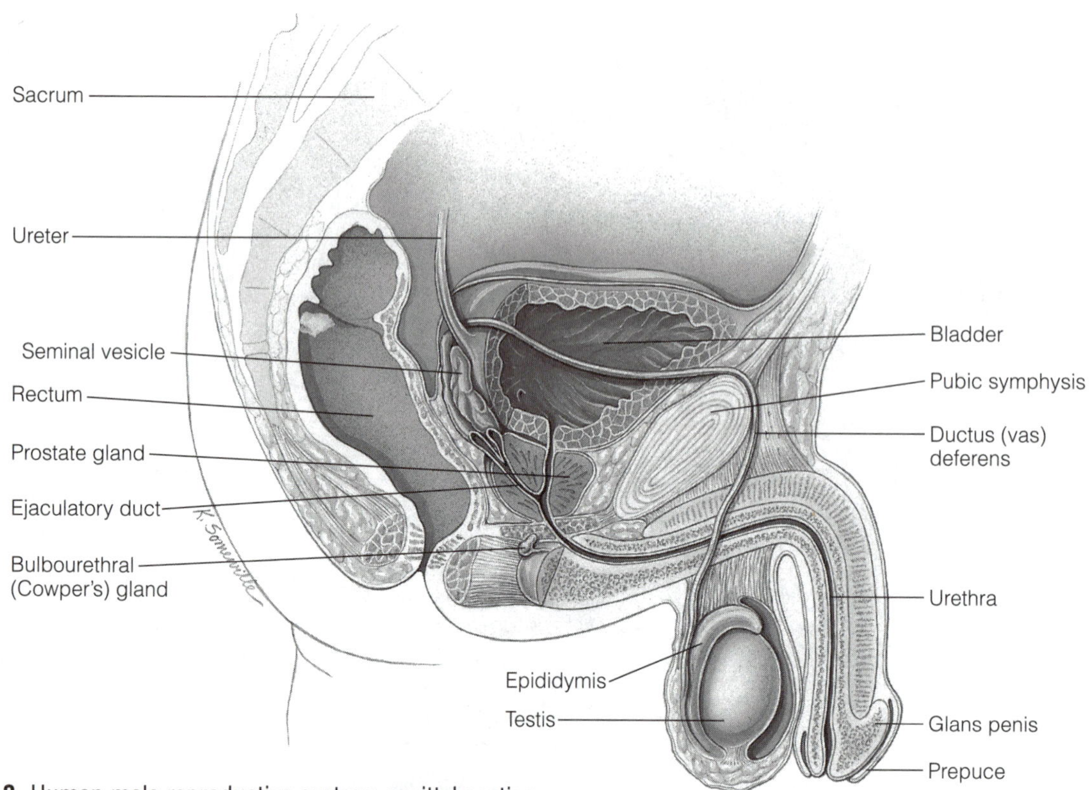

Figure 17.3 Human male reproductive system: sagittal section.

After sperm have been produced in the seminiferous tubules of the testes, they are stored in a highly coiled tube, the **epididymus**, located at the rear of the testis (Figure 17.3). From here they enter the **vas deferens** (sperm duct), which is about 18 inches long and passes through an opening in the body wall called an **inguinal canal**. This is the place where the testis descended before birth in the groin region. An inguinal hernia involves a part of the intestine protruding through the inguinal canal—a painful but operable condition.

A **vasectomy**, a surgical procedure that provides birth control in males, can be performed by cutting and tying off both ends of the vas deferens. This causes no change in sex characteristics or sexual desire, but it does prevent sperm from being available for fertilization.

Each vas deferens receives the duct from a seminal vesicle, forming the **ejaculatory duct** (Figure 17.4). The right and left ejaculatory ducts empty into the urethra. The **prostate gland** surrounds the urethra at this point.

Seminal vesicles are glands that produce a sugary fluid (fructose), which provides energy for sperm.

The muscular/glandular **prostate gland** produces a milky alkaline fluid which is believed to activate sperm motility. In older men the prostate sometimes enlarges and interferes with urination as it puts pressure on the urethra.

Cowper's or **bulbourethral** glands are two pea-sized, yellowish glands located at the base of the penis. Their secretion is a clear mucuslike fluid that flows during sexual stimulation and empties into the urethra. It lubricates the glans penis and urethra and may help neutralize the acidity of the urethra.

The **penis** is the male copulatory organ. It contains **erectile tissue** composed of sinuses that fill with blood and cause erection during sexual stimulation. Violent peristalsis of the ejaculatory ducts and urethra discharge the **seminal fluid** (semen) during sexual stimulation and intercourse.

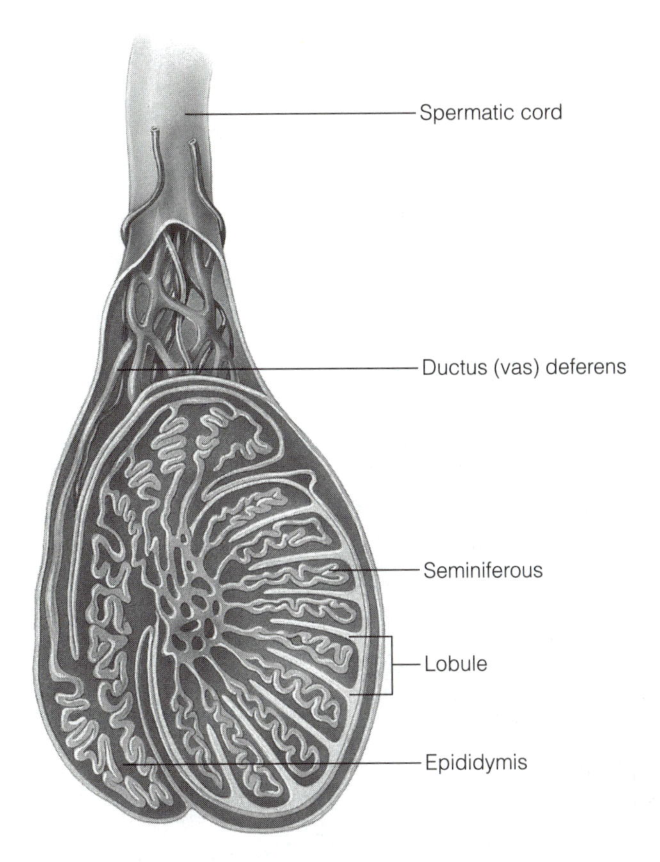

Figure 17.4 Human testis—duct system.

Seminal emission averages 3–5 ml (a teaspoonful) and contains an average of 300–400 million sperm, although the range is 45–730 million. Less than 35 million per ejaculation is considered sterility.

Each sperm consists of three major parts: a **head**, **midpiece**, and **tail**. The head contains the nucleus of the cell it was formed from and contains one-half the chromosome number of the male that produced it. The midpiece contains mitochondria and ATP. The tail (or flagellum) moves rapidly to propel it through the uterus and the fallopian tubes (oviducts) of the female.

ACTIVITY: Studying the Male Reproductive System

1. Using your microscope, observe the prepared slide of human testis cross-section. Use Figure 17.1 on page 157 as a guide. Locate the seminiferous tubules and mature sperm cells.
2. Observe the prepared slide of a human sperm smear. Use high power. Locate the three major parts of the sperm cell: head, midpiece, and tail.
3. Observe the human torso model and other available models. Review all of the male reproductive structures you have learned today.

Male Reproductive System of the Fetal Pig

Note: Although you will dissect only your own specimen, you are responsible for knowing the reproductive structures of both sexes. When you are finished with your dissection, you should exchange your specimen with one of the opposite sex dissected by a fellow student. Use Figure 17.5 as a reference.

1. Locate the urogenital opening just posterior to the umbilical cord and the immature scrotal sacs ventral and lateral to the anus. Using scissors, carefully cut through the scrotal sac only. Free the testicle from the sac by cutting the attaching tissue. Cut the membranous sac. Observe the testis and the posterior lateral structure, the epididymis.
2. Starting at its opening posterior to the umbilical cord, separate the penis from the ventral body wall and trace it to the point at which it goes behind (dorsal to) the pelvic bone, which you can locate by its hardness. Locate each sperm duct or vas deferens where each loops over the ureter, and follow them toward the testes. In the younger specimens the testes are elliptical-shaped bodies that may be found in the abdominal cavity. In the older ones, they are already in the scrotal sac.

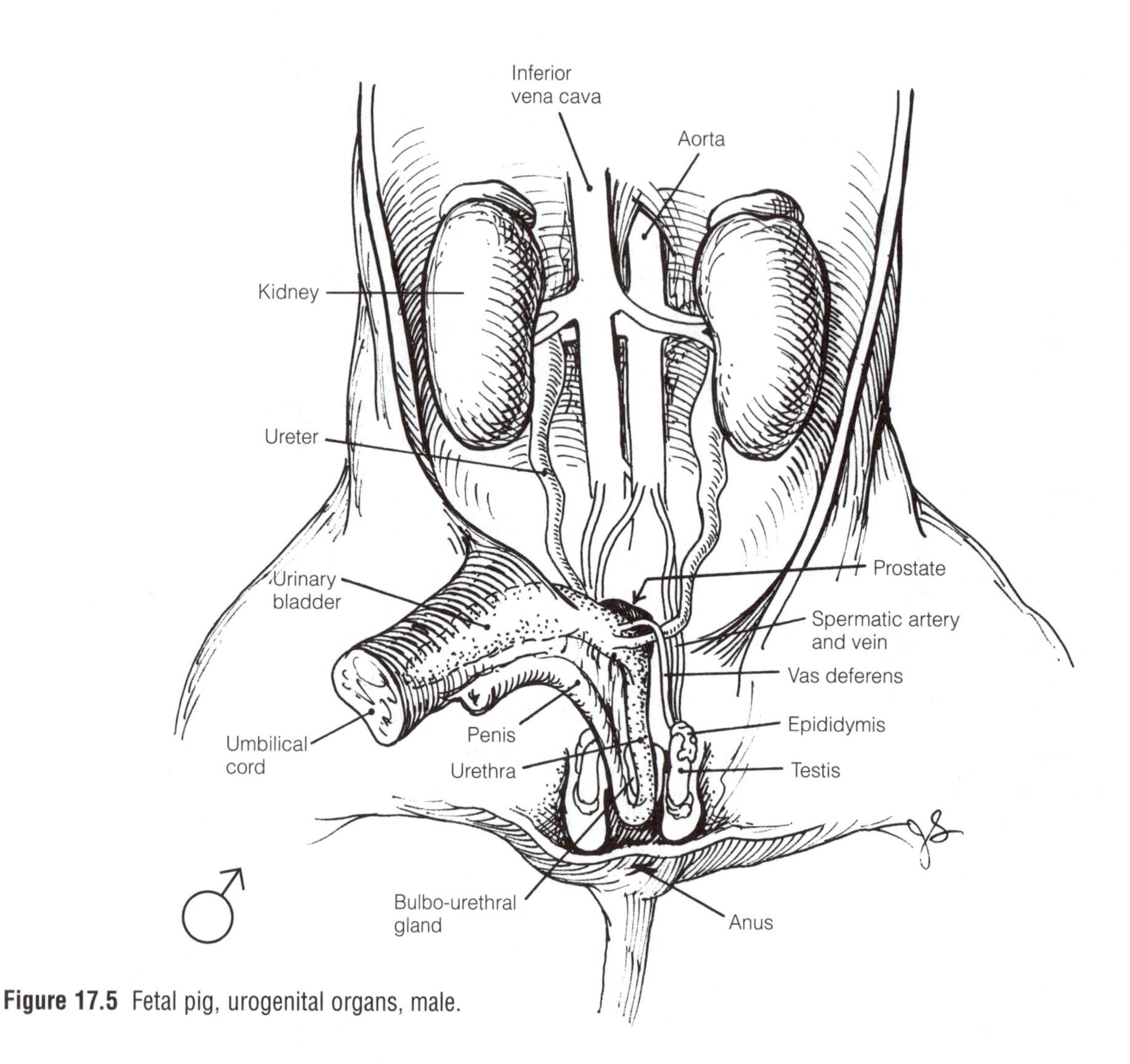

Figure 17.5 Fetal pig, urogenital organs, male.

3. Follow the vas deferens through the inguinal canal, which connects the abdominal cavity and the cavity of the scrotal sac, and through the ventral surface of the scrotal sac. Near the testes, the vas deferens connects with the epididymis, a crescent-shaped organ enveloping the posterolateral surface of the testes. Running through the inguinal canal, alongside the vas deferens, are the spermatic blood vessels and nerves.

4. To follow the urethra from the bladder to the penis, it is necessary to cut through the pelvic girdle. Spread the hind legs apart and remove the skin and muscle lying ventral to the pelvic girdle. This should expose a thin white line, the pubic symphysis. Insert the scalpel into the pubic symphysis and exert enough pressure to split the girdle in two. Be careful not to cut into the urethra, which lies dorsal to the symphysis. Find the Cowper's glands (long white bodies on either side of the urethra), and the seminal vesicles (small glands dorsal to the region where the vas deferens join the urethra). The prostate gland is too poorly developed in the fetal animal to locate. ■

Study Questions

1. List the three major accessory glands of the male reproductive system. Give their functions.

   a. ______

   b. ______

   c. ______

2. Trace the pathway of a sperm cell from the time it is formed until it leaves the male body: Name all the structures it passes through in proper order.

3. What is castration? What are its results in male animals such as altered cats, capons, geldings, and eunuchs?

4. What is testosterone? List some of its major functions.

5. Sketch a human sperm cell with its three major parts. List the functions and contents of each part.

6. What cell division process is involved in spermatogenesis? Why is this type of cell division essential for sexual reproduction?

7. What are the advantages of sexual reproduction?

EXERCISE

18 The Female Reproductive System

OBJECTIVES	MATERIALS
After completing this exercise, you should be able to: 1. Name the internal and external female reproductive organs and know their location in the body. 2. Understand the relationship between the structure and function of the female reproductive system.	❑ Models of female reproductive organs (1 per table) ❑ Pelvic torso—female reproductive organs (demo) ❑ Ovary (demo) ❑ Male and female pelvic girdles (demo) ❑ Pregnant uterus series (demo) ❑ Female breast (demo) Prepared slides: ❑ Mammalian uterus, fetus (demo) ❑ Mature ovary, Graafian follicle

Introduction

In addition to providing reproductive cells, the female body is specially adapted to support development of the fetus and nourish the newborn.

Anatomy of the Female Reproductive System

The **ovary** is the primary reproductive organ of the female. The ovary functions as both an endocrine gland (hormones) and an exocrine gland (ova).

The **accessory structures** of the female reproductive system transport, house, and nurture the reproductive cells and the developing fetus.

For ease of study, the female reproductive system is discussed in terms of two divisions—the internal organs and the external organs.

External Female Genitalia

As you read the following description, familiarize yourself with the structures described by referring to Figures 18.1 and 18.2.

The external genitalia, or **vulva**, consist of the following:

- **Mon pubis** a rounded fatty and fibrous tissue mound which overlies the pubic symphysis. During puberty, when the secondary sexual characteristics develop, it becomes covered with coarse hair.
- **Labia majora** two elongated, pigmented, hair-covered skin folds running inferiorly and posteriorly from the mons pubis. Its homologue (similar in embryonic origin) in the male is the scrotum.
- **Labia minora** two smaller, hair-free folds enclosed by the labia majora.

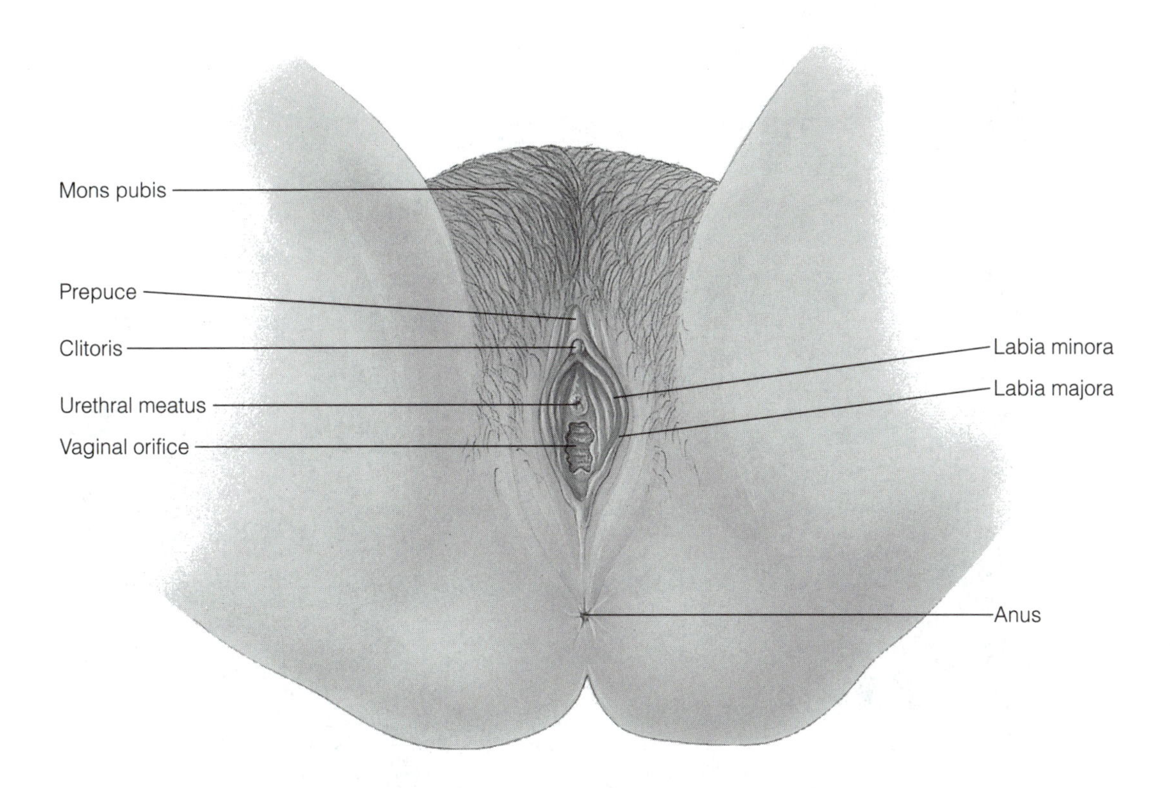

Figure 18.1 Human female reproductive system: external genitalia.

- **Clitoris** a small, protruding, highly sensitive structure composed of erectile tissue. Its homologue is the male penis. While in males the urethra is contained within the penis, in females the urethra is *not* contained within the clitoris. The clitoris is hooded by the anterior meeting of the labia minora. This hood is referred to as the prepuce of the clitoris.
- **Vestibule** the region enclosed by the labia minora.
- **Urethral orifice** the opening of the urethra. Part of the excretory system, the urethral orifice drains the bladder. It is located within the vestibule posterior to the clitoris and anterior to the vaginal orifice.
- **Vaginal orifice** the opening into the vagina. It is formed within the vestibule just posterior to the urethral orifice. You will notice that in the female, the openings of the urinary and reproductive systems are separate structures.
- **Bartholin's glands/vestibular glands** mucus-secreting glands on either side of the vaginal opening whose secretions lubricate the distal end of the vagina during intercourse (these glands are not illustrated in Figure 18.1). Their male homologue is the bulbourethral (Cowper's) gland.
- **Hymen** a thin fold of mucous membrane which may partially or fully close the vaginal opening.
- **Mammary glands** while not part of the vulva, the mammary glands are included here as an external reproductive organ. These modified sweat glands exist in both sexes but have a reproductive function only in the female. Their function is to produce milk to nourish the newborn infant. This milk production is stimulated by hormones only when a birth actually takes place.

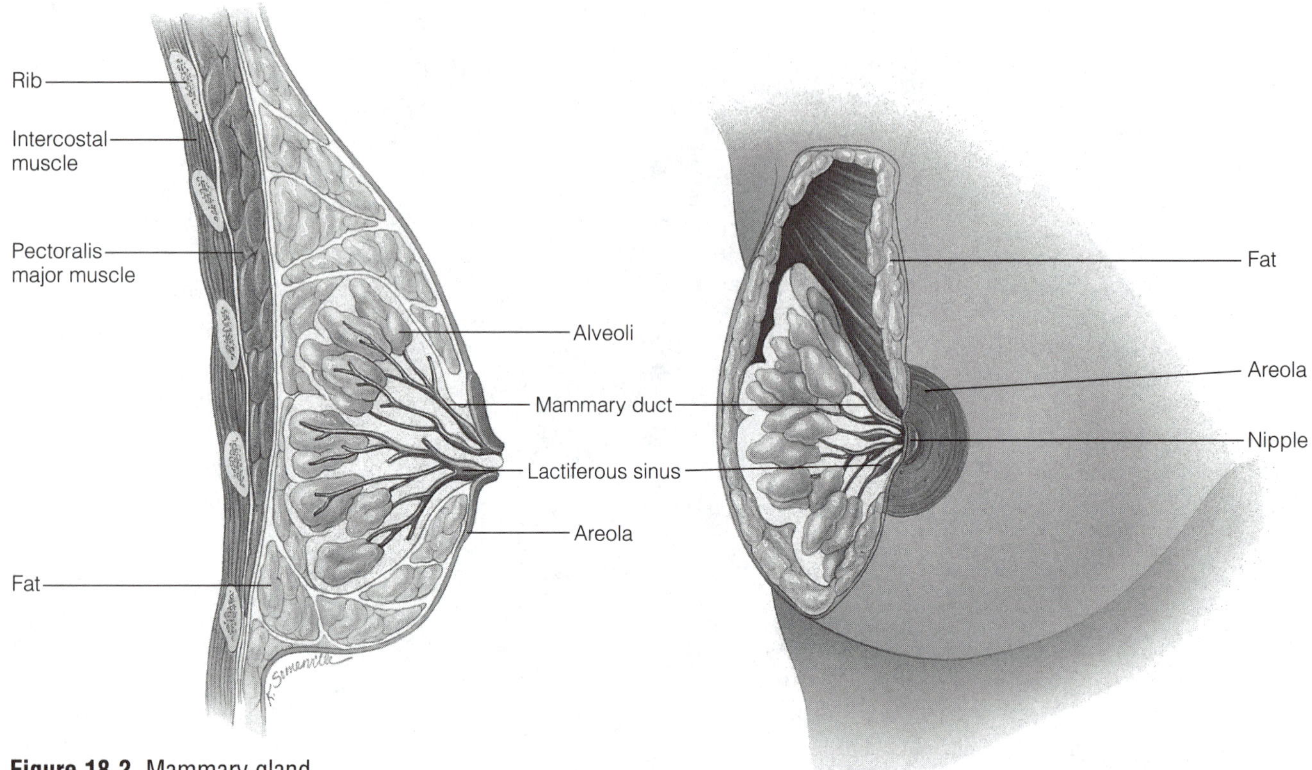

Figure 18.2 Mammary gland.

Internal Female Reproductive Organs

As you read the following descriptions, familiarize yourself with the structures described by referring to Figures 18.3 and 18.4 and the model at your table.

- **Ovaries/female gonads** the primary reproductive organs are paired, almond-shaped glands positioned in the upper pelvic cavity and lying near the ends of the fallopian tubes. They are held in position by several ligaments. Beneath the outer ovarian epithelium is a layer containing the developing ova and surrounding follicular cells. These developing ova are inactive until puberty. After puberty, pituitary hormone stimulation causes growth of follicles.
- **Fallopian or uterine tubes** the two tubes that enter the upper region of the uterus. The distal ends of the tubes are funnel-shaped and have fingerlike projections called **fimbriae**. There is no actual contact between the ovary and fimbriated ends of the uterine tubes. The waving ends of the fimbriae create a current that sweeps the egg from the ovary into the fallopian tube, where the egg is propelled toward the uterus by the cilia of the tubule walls.
- **Uterus** a pear-shaped, highly muscular organ situated between the bladder and the rectum. Its narrow end, the **cervix**, is directed downward. The major portion of the uterus is referred to as the **body**, and its upper rounded region, above the entrance of the fallopian or uterine tubes, is called the **fundus**.
- **Vagina** a muscular tube that extends approximately 15 cm from its opening in the vestibule to the uterus above. It serves as the female copulatory organ and the birth canal. It also permits the passage of the menstrual flow from the uterus to the exterior.

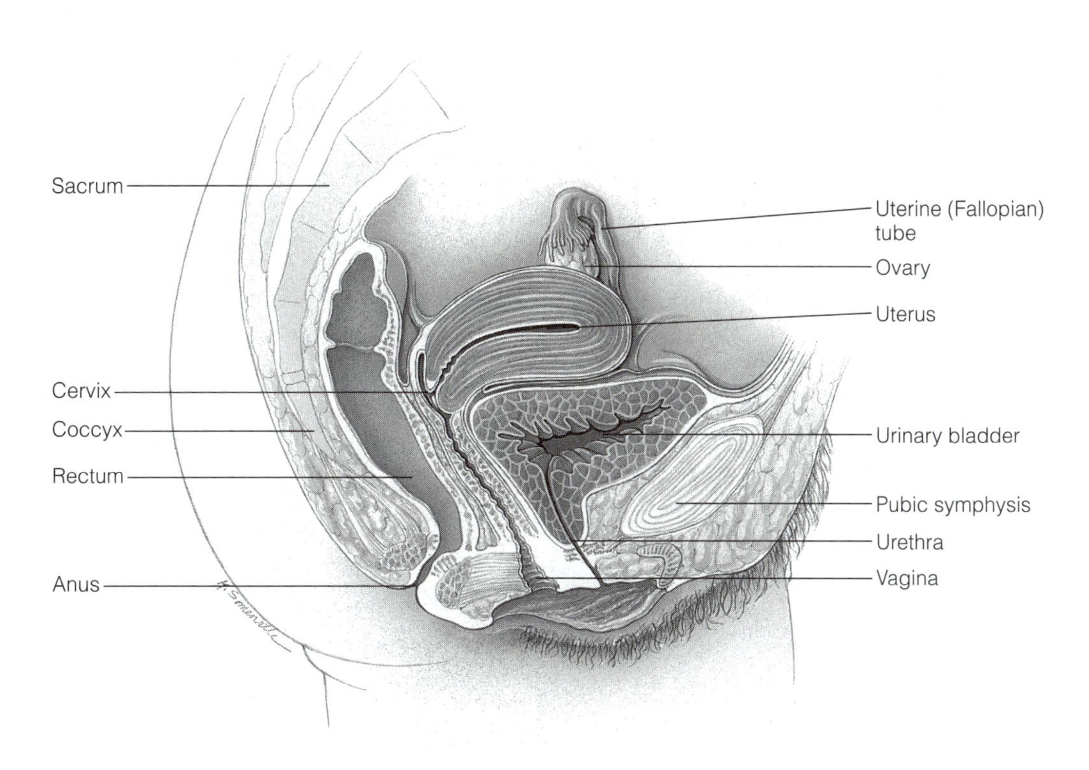

Figure 18.3 Human female reproductive system: mid-sagittal section.

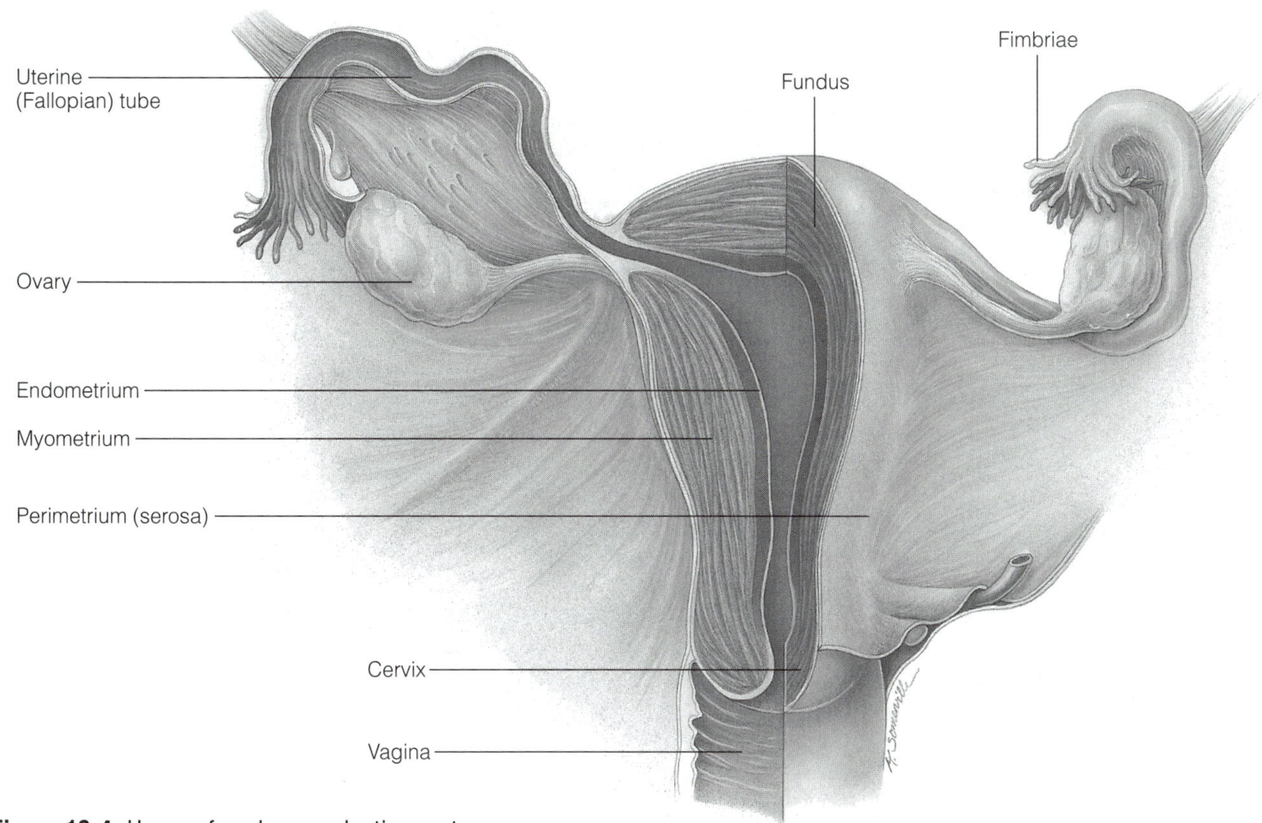

Figure 18.4 Human female reproductive system.

ACTIVITY: Studying the Female Reproductive System

1. Using the preceding diagrams (Figures 18.3 and 18.4), locate all of the labeled structures on the models and demonstration slides.
2. Observe the individual internal organ models and identify the structures listed above.
3. Observe the pelvic model of the female on the side table and make special note of the position of the female organs in relationship to the other pelvic organs.

Reproductive System of the Female Fetal Pig

Refer to Figure 18.5 while performing this activity.

1. Find the paired, pale, oval ovaries lying in the abdominal cavity located a short distance caudal to the kidneys. Surrounding the edges of each ovary is a very narrow, convoluted fallopian tube. The fallopian tube is very short; it leads into the horns of the uterus. These two horns converge to the midline of the animal and there unite to form the body of the uterus. Most animals carry multiple fetuses in the uterine horns not in the body of the uterus. This enlarged uterus enables the pig to produce litters. The body of the uterus extends posteriorly for a short distance and ends in the neck, or cervix, which opens into the vagina.

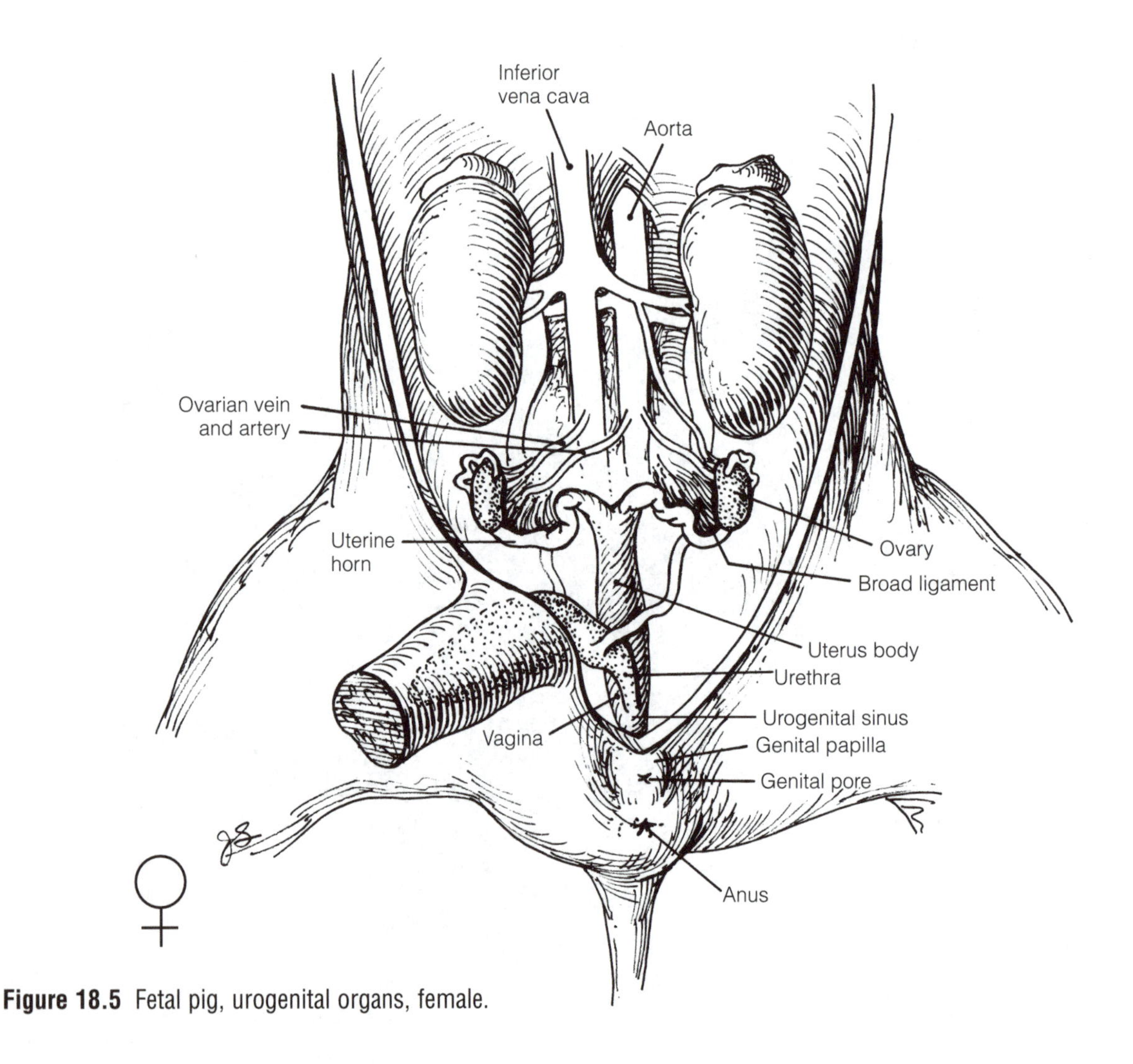

Figure 18.5 Fetal pig, urogenital organs, female.

2. To dissect the vagina and to find its connection with the urethra, it is necessary to cut through the pelvic girdle. You can locate the pelvic girdle by its hardness. Spread the legs apart and remove the skin and muscles ventral to the girdle. This will expose a thin white line, the pubic symphysis.

3. Insert the scalpel into the pubic symphysis and exert just enough pressure to split the girdle in two, but be careful not to cut into the urethra, which lies dorsal to the symphysis.

4. Separate the urethra from the vagina to the point where they fuse to become the urogenital sinus. Follow the sinus to its external opening. On the ventral edge of the opening is the urogenital papilla. ■

Study Questions

1. Why is a sterilization operation that involves severing or ligating the tubes more dangerous for a woman than for a man?

____

____

____

____

2. List the female reproductive accessory organs.

____

____

____

____

3. Give two reasons why a woman must urinate more frequently when pregnant.

____

____

____

____

4. Trace the pathway of an egg from its origin in the ovary until it is expelled from the body either unfertilized or as a child.

____

____

____

____

____

____

____

____

INDEX